FEMINIST EPISTEMOLOGY AND PHILOSOPHY OF SCIENCE

W0113676

Feminist Epistemology and Philosophy of Science: An Introduction is structured around six questions and the answers to them that have been offered by feminist epistemologists and philosophers of science. By showing how these answers differ from those of traditional philosophical approaches, the book situates feminist work in relation to philosophy more generally.

The questions are: Who knows? What do we have knowledge of? How do we know? What don't we know? Why does it matter? and How can we know better? In addressing these questions, the book reviews feminist accounts of objectivity, agnotology, issues in social epistemology – including epistemic injustice – and considers how feminist epistemology and philosophy of science aim at better knowledge production. The audience for the book is upper division undergraduates, but it will be useful as a foundation for graduate students and other philosophers who are seeking a general understanding of feminist work in these areas.

Key features:

- Provides an overview of contemporary feminist epistemology and philosophy of science
- Contrasts feminist epistemology and philosophy of science with traditional philosophy in these areas
- Provides clear examples of the benefits of feminist approaches
- Includes in each chapter an initial overview and, at the end of the chapter, suggested additional readings and discussion questions.

Sharon Crasnow is Distinguished Professor Emerita at Norco College in Southern California, USA. Her main area of research is the relationship between methodology and epistemology in the sciences. She is co-editor of *The Routledge Handbook of Feminist Philosophy of Science* (with Kristen Intemann).

Kristen Intemann is a Professor of Philosophy in the Department of History and Philosophy and Director for the Center for Science, Technology, Ethics and Society at Montana State University, USA. Her research and teaching interests include feminist philosophy and philosophy of science, particularly issues of objectivity, values in science, and epistemic trust.

FEMINIST EPISTEMOLOGY AND PHILOSOPHY OF SCIENCE

An Introduction

Sharon Crasnow and Kristen Intemann

Routledge
Taylor & Francis Group

NEW YORK AND LONDON

First published 2024
by Routledge
605 Third Avenue, New York, NY 10158

and by Routledge
4 Park Square, Milton Park, Abingdon, Oxon OX14 4RN

Routledge is an imprint of the Taylor & Francis Group, an informa business

© 2024 Sharon Crasnow and Kristen Intemann

British Library Cataloguing in Publication Data
A catalogue record for this book is available from the British Library

Library of Congress Cataloging-in-Publication Data
A catalog record has been requested for this book

ISBN: 978-1-032-69376-7 (hbk)
ISBN: 978-1-032-69374-3 (pbk)
ISBN: 978-1-032-69378-1 (ebk)

DOI: 10.4324/9781032693781

Typeset in Sabon
by Taylor & Francis Books

CONTENTS

To all of those who persevered to remove obstacles for others in fields where they were underrepresented and to those who pursued feminist philosophy, even when it involved professional risks.

PREFACE

Background

The impetus for this book was our work on *The Routledge Handbook of Feminist Philosophy of Science* published in 2021 (and the urging of our editor at Routledge, Andrew Beck), but the source of our collaboration goes back to our meeting 20 years ago. We were both participants in the 2003 NEH Summer Institute on Science and Values at the University of Pittsburgh. The Institute was led by Peter Machamer and Sandra Mitchell and, although it was not specifically focused on feminist perspectives, several of the participants were. Those among us with a feminist bent formed a reading group and both the Institute and that reading group proved to be the origins of collaborative, long-lasting friendships that have sustained us intellectually and otherwise throughout the past 20 years. We and this book owe much to that Summer Institute and to our fellow travelers – Evelyn Brister, Sharyn Clough, and Inmaculada de Melo-Martín.

Why did we write the book?

The more academically motivated roots of the book lie in a felt need for a more current overview of feminist epistemology and philosophy of science. Other available books, many of which are still useful, come from an earlier period in feminist thought. Linda Alcoff and Elizabeth Potter's edited collection, *Feminist Epistemologies* appeared in 1993, for example. Alessandra Tanesini's *Introduction to Feminist Epistemology* is from 1999. Potter's *Feminism and Philosophy of Science: An Introduction* is somewhat more recent (2006). The collection edited by Heidi Grasswick *Feminist*

Epistemology and Philosophy of Science: Power in Knowledge (2011) is excellent and individual papers in the volume give a good sense of how feminist philosophy of science and epistemology have developed since the 1990s. However, we felt that there was still room for a book that offers the historical context out of which feminist epistemology and philosophy of science arose and that situates current work both in relation to that history and to contemporary mainstream epistemology and philosophy of science.

We would not be practicing feminist epistemology if we did not acknowledge that this is very much *our* take on how feminist epistemology and philosophy of science are historically and intellectually situated. As a result, there are gaps in our treatment. For example, we have not covered every feminist approach to the question of objectivity but have focused on those that we think are most promising and with which we are most familiar. We also acknowledge that the approaches we focus on are primarily in the Anglo-American tradition and that the feminist approaches considered are contrasted with primarily Western analytic approaches to epistemology and philosophy of science. While incomplete with respect to some areas and topics, we nonetheless believe the book provides a good introduction and we hope that it will encourage readers to continue to explore the various topics covered here. We have suggested some additional readings to aid in that goal.

Who is the book for?

We imagine our primary audience to be upper division undergraduate students in a class devoted to feminist epistemology or philosophy of science. Much of the available feminist work in these areas is more appropriate for an audience that already has some familiarity with feminist literature. We hope that this text will provide the sort of foundation that would be helpful in more easily accessing that literature. We also intend the book to provide an overview that graduate students can build on. The additional suggested readings at the end of each chapter should aid in this goal as well. Our references also provide other resources that instructors might use to augment the text. Finally, we hope that any philosopher who wants to incorporate more feminist work into their mainstream philosophy of science or epistemology course, or would like to know more about feminist epistemology and philosophy of science should be able to make use of the book for those purposes.

How can the book be used?

The six questions around which the book is organized came to us before we even started to write. We see these as fundamental questions about knowledge and have characterized feminist philosophy as answering them differently than much mainstream philosophy had. At the same time, we point out that feminist philosophers have not been the only ones to seek alternative approaches in epistemology and philosophy of science, although they were at the forefront of doing so and the contributions they have made are not always fully appreciated. Thus, another goal has been to show how feminist work and contemporary mainstream epistemology and philosophy of science are more closely integrated than is sometimes acknowledged.

Although our intention has been to write a book that presents a coherent picture of feminist epistemology and philosophy of science, in part, by showing the connections among the answers to the six questions we have posed, it is nonetheless possible to pick and choose among the chapters depending on the focus of the course being taught or the interests of the reader. Chapter 1, the introduction, provides a summary of the book as a whole and serves as an overview of the issues. Chapter 2 focuses on the knower and might be useful in relation to a course that focuses on questions of social ontology and identity and so could be used outside of an epistemology or philosophy of science course. Chapters 3 and 4 each explore different aspects of objectivity and the role of values in knowledge and science. Chapter 3 explores different ways in which knowledge might be objective and focuses on how feminists have changed what it is that we know. Chapter 4 covers many of the standard issues that have arisen when feminist philosophers have attempted to address worries about values in science and epistemological relativism and gives accounts of several of the main approaches feminists have taken to addressing worries about relativism. Chapter 5 could stand alone as an introduction to and summary of the agnotology literature. Chapter 6 serves a similar function for epistemic injustice and might also stand alone or could be used in conjunction with many of the readings referenced in that chapter. Although we believe the book as a whole has an arc culminating in Chapter 7, the separate chapters may prove useful in various contexts.

A good companion to the text is our edited volume, *The Routledge Handbook of Feminist Philosophy of Science* (2021). Although we have referenced chapters from the handbook in the text, the following is a list of the most closely aligned chapters of the handbook for each chapter of this book:

Chapter 1: Chapter 1 The Origins of Philosophy and Science in Ancient Greece; Chapter 20 Feminist Philosophy of Biology; Chapter 21 Observing Primates
Chapter 2: Chapter 2 Margaret Cavendish and the New Science; Chapter 3 Emily Du Châtelet; Chapter 4 The Rocket Women of India; Chapter 5 Contributions of Women to Philosophy of Science; Chapter 28 Feminist

Methodology in the Social Sciences; Chapter 30 What Is It Like to Be a Women in Philosophy of Physics?

Chapter 3: Chapter 14 Is Sex Socially Constructed?; Chapter 15 Feminist Perspectives on Values in Science; Chapter 16 Situated Knowledge and Objectivity; Chapter 18 How the Facts Might Give Us Socially Responsible Science; Chapter 27 Feminist Economics; Chapter 29 Feminist Approaches to Concepts and Conceptualization

Chapter 4: Chapter 6 Feminist Empiricism; Chapter 7 Thinking Outside-In; Chapter 13 Where Are All the Pragmatist Feminist Philosophers of Science?; Chapter 15 Feminist Perspectives on Values in Science; Chapter 16 Situated Knowledge and Objectivity; Chapter 28 Feminist Methodology in the Social Sciences

Chapter 5: Chapter 8 Latin American Decolonial Feminist Philosophy of Knowledge Production; Chapter 17 Ignorance, Science, and Feminism

Chapter 6: Chapter 9 Sciences of Consent: Indigenous Knowledge and Governance, Value, and Responsibility; Chapter 10 Queer Science Studies/Queer Science; Chapter 11 Naturalizing and Denaturalizing Impairment and Disability in Philosophy and Feminist Philosophy of Science; Chapter 12 Epistemic Vices and Feminist Philosophy of Science

Chapter 7: Chapter 18 How the Facts Might Give Us Socially Responsible Science; Chapter 19 Feminist Science for the People; Chapter 26 Implementing Intersectionality in Psychology with Quantitative Methods; Chapter 31 Inclusivity in Engineering Education

Who do we acknowledge?

A central thesis of feminist epistemology explored in this book is that it takes communities to produce knowledge. Our own understanding and work as feminist scholars has benefited from and developed through the work of other feminist scholars who have contributed to our thinking and understanding of feminism, feminist epistemology, and feminist philosophy of science. These influences are reflected throughout the book but there are others who have also encouraged us, mentored us, taught us, had long conversations with us, or otherwise supported us who we would like to mention specifically. These include: Matthew J. Brown, Ann Garry, Heidi Grasswick, Sandra Harding, Janet Kourany, Helen Longino, Nancy McHugh, Lynn Hankinson Nelson, Elizabeth Potter, Sarah Richardson, Kristina Rolin, Phyllis Rooney, Miriam Solomon, Joanne Waugh, and Alison Wylie. We are surely missing others here because there have been so many over the years. In addition, there are those who have served as mentors, supported our work, and influenced the fields discussed here even though they do not explicitly work in feminist theory or feminist philosophy. These include Nancy

Cartwright, Arthur Fine, Hugh Lacey, Mary Morgan, Kristin Shrader-Frechette, and Andrea Woody. We would also like to give special thanks to Melissa Jacquart, who provided helpful feedback on the early chapters of this manuscript and to all of those who contributed to the *Routledge Handbook on Feminist Philosophy of Science*, who helped us realize the breadth and depth of new work being done in this field. We also acknowledge two anonymous reviewers, both of whom read the text carefully and provided invaluable, constructive comments. Last, but not least, we would like to thank our families who supported us while we traveled, discussed, and Zoomed during our work on this project.

1

WHAT IS FEMINISM AND WHAT DOES IT HAVE TO DO WITH SCIENCE?

Overview

This introductory chapter clarifies how feminism, epistemology, and philosophy of science are understood in the text. The account of feminism offered focuses on addressing oppression that results from the unequal distribution of power due to gender but also notes that gender intersects with other social categories that affect participation in knowledge production in a variety of ways. The chapter includes a brief history of feminist epistemology and philosophy of science and provides an overview of how feminist approaches differ from traditional epistemology. While other philosophers and historians have challenged traditional views since the mid-20th century, feminists have made significant contributions to alternate understandings of the production of knowledge (including scientific knowledge). The six questions addressed in the subsequent chapters provide an account of how and why this is so.

Introduction

In the 1970s and 1980s, more women began entering scientific fields from which they had previously been largely excluded. In some cases, women scientists made surprising discoveries. For example, women who entered the field of primatology made observations that appeared to conflict with long-standing assumptions about potential differences in male and female primates (Hrdy 1986). The traditional view, which went in hand with the understanding of evolutionary theory at the time, was that males were "promiscuous" (having multiple reproductive partners) and females were "coy", picky, and monogamous (tending to have one reproductive partner). This

DOI: 10.4324/9781032693781-1

was thought to be both a function of biology (female primates only have one offspring at a time, whereas males could potentially have multiple offspring at once) and also evolutionary fitness (males that have more offspring are more likely to have some offspring survive while a female must ensure that the male they mate with is likely to invest in the protection and survival of that offspring). Yet women primatologists observing the behaviors of female primates quickly discovered that female primates were often just as promiscuous as their male counterparts, and later developed theories that such behavior would be evolutionarily advantageous because the more males who might become invested in the care and protection of offspring the more likely the offspring would survive.

This kind of discovery, however, challenged not only widespread assumptions about primates, sex differences, and roles in evolution. It also challenged fundamental assumptions in epistemology and philosophy of science. For centuries, philosophers had assumed that the social features of scientists or knowers were not relevant to what we know or what evidence we have. Insofar as humans have essentially the same perceptual apparatus and ability to reason, what is observed should not depend on contingent features of an observer. Moreover, scientific methods were believed to be "objective" so that, in theory, whatever data or evidence was collected should not depend on who was collecting it.

What then could explain that women primatologists appeared to observe things that were not previously observed by other scientists and were relevant to understanding primate behavior? It is quite dubious that women scientists are somehow just *better* observers! How is it that they observed something different? Was it that they asked different questions? Did they set out to observe different things? Did they use different methodologies? Were they more open to questioning the status quo? The answers to these questions may raise other, more fundamental questions, for the philosophical study of knowledge and our understanding of science. Are there features of knowers that matter for what is known or how it is known? Is science "objective" and what does that mean? How can we identify and remove bias from the production of knowledge? These sorts of questions gave rise to feminist epistemology and philosophy of science.

What exactly are feminist epistemology and philosophy of science and why are they of interest? A goal of this book is to answer these questions. In order to do so, we examine some basic presuppositions that have directed epistemology and philosophy of science more generally – questions that feminist philosophers have given different answers to than philosophers had offered in the past. We use six questions to shape the account we provide in this book. In addressing these questions, we aim to bring about a more complete picture of both what feminist epistemology and philosophy of science set out to do and the ways that such approaches may make for better understanding of knowledge production. The questions we address are the following:

- Who knows?
- What do we have knowledge of?
- How do we know?
- What don't we know?
- Why does it matter what we know?
- How can we know better?

Why these questions are important to ask (and to answer) is best seen by contrasting feminist approaches to traditional approaches to epistemology and philosophy of science. While each question will be addressed in Chapters 2–7, this introductory chapter offers an overview. First, however, we lay some groundwork by clarifying how we understand feminism, epistemology, philosophy of science, and their intersections.

Feminism

There are a many different versions of feminism, but in seeking to identify a shared understanding of the term, we have settled on the following. Feminism is an emancipatory political movement with the aim of ending oppressive structures based on gender ideologies and norms that divide the world into two classes of people with distinct social roles—men and women—that tend to systematically empower men and disempower women. Although there is disagreement among feminists about how to bring about the desired result, this core aim is generally agreed on. Disagreements tend to center on the question of how deeply the prevailing ideologies, structures, and institutions of society empower men and disempower women and what are the best ways to address that. We take our characterization to be a minimalist one and, as we delve into the various ways in which feminist epistemologists and philosophers of science have interpreted feminism, our understanding of the term will expand.

For now, it is worth noting that even this minimalist understanding of feminist goals requires several additional commitments. First, feminism derives in part from the recognition that power is distributed unequally, and that gender is one of the social factors affecting its unequal distribution. Second, feminists are committed to an examination of the various ways in which societal structures and institutions are shaped through this differential distribution of power. Third, feminism is at its core a political movement committed to the amelioration of structures that hinder or prevent the achievement of egalitarian goals. Finally, although it may not be obvious from the initial characterization, our understanding of feminism recognizes that ending the oppression of women is inseparable from ending all forms of oppression more generally. This is true in part because women are among the members of all races, ethnicities, religions, sexualities, and abilities and so their liberation as women is never wholly separable from the effects of being subject to these other intersecting identities. It may also be the case that the ways in which sexist oppression manifests and operates is different for those with multiple intersecting identities. For example, stereotypes of femininity and gender norms may intersect with race or class stereotypes. What counts as "feminine" may be different for white women and women of color, or for women of different socioeconomic classes. Additionally, it may turn out that the mechanisms or causes of different forms of oppression are intersecting and interrelated. For example, some feminists have argued that capitalism is a system that is at the root cause of gender and racial oppression as well as class oppression. Some have also pointed to ways in which sexist stereotypes are enforced and maintained through heteronormativity or norms for sexuality. Thus, our minimalist account of feminism is intended to be inclusive and to create space for theorizing the intersectionality of identities and systems of oppression.

Before leaving this brief discussion of feminism, we need to address the issue of how to understand the claim that feminism seeks equal rights and opportunities for *women*. By "women" we mean simply those who identify as women or are perceived by others as women and are subject to oppression as a result. We offer no further analyses here, although we do specifically reject analyses that depend on essentialist biological identifications.

Epistemology and philosophy of science

Philosophy of science and epistemology are closely related to each other. Epistemology is the study of knowledge, including the difference between knowledge and other cognitive states. More specifically, epistemology has traditionally included the study of belief and its relationship to knowledge. Science is often taken to be the paradigm case of knowledge and so epistemology and philosophy of science overlap. However, there are a variety of other philosophical topics of interest for philosophers of science, such as: understanding key concepts used in science (for example, species for biology, space-time for physics), examining key elements of scientific reasoning, such as evidence, confirmation, theory acceptance, and scientific explanation. Philosophy of science also explores the aims of science and includes debates about those aims. Other topics of interest to philosophers of science are the relationship between science and society, the use of scientific expertise to guide policy decisions, and ethical issues having to do with both scientific research and the application of the results of such research.

From the account given so far, it may not be obvious why or how feminism intersects with epistemology or philosophy of science. Questions about ethical issues and the relationship between science and society might be areas of contact but the most clearly epistemological questions – the relationship between knowledge and belief, the nature of evidence or confirmation – might seem far removed from the goals of feminism. Thus, it may be unsurprising to learn that when feminist epistemology and philosophy of science began to develop in the 1980s and into the 1990s it was not warmly welcomed by traditional epistemologists or philosophers of science. At best, feminism was thought to be irrelevant to these areas of philosophy. At worst, it was thought be incompatible with or antithetical to understanding the objectivity of science or establishing objective criteria for knowledge.

A brief history of the subfield will help to clarify both feminist epistemology and philosophy of science, but the full picture will only emerge in Chapters 2–7 as we explore the six questions that we have posed.

A history

The short answer to the question of how feminism came to be seen as relevant to questions of knowledge is that during the second half of the 20th century, there was a shift in how knowledge, including scientific knowledge, was understood. One of the primary features of this shift was the recognition of the role of social, cultural, and historical contexts in the production of knowledge and their effect on the knowledge produced.

Awareness of the social situatedness of knowledge arose from a variety of sources within philosophy of science but also from within science itself. The increasing participation of women in public life – the world of work outside the home, including scientific work – both shifted and resulted in the shifting of the social milieu. Women who found themselves hindered or even blocked from participating in academia or the sciences, pushed through, or resisted these harms, through their own courageous norm-breaking, the support of sympathetic men within the profession, or ultimately through securing greater access to knowledge production through legal interventions. When women were able to enter fields not previously open to them, they often found that it was difficult to study subjects they were interested in and in the ways that they thought best suited to do so, because both the subjects that interested them and the approaches they took to investigate these subjects were unlike or even antithetical to those most established in their professions. This was likely not because of any biological difference, but because they had been socialized under certain social, political, and economic structures that had produced different experiences, interests, and skill sets.

While this social shift began happening for women entering scientific and other academic fields, there were also transformations happening in philosophy of science that acknowledged the social situatedness of knowledge. Within the philosophy of science, the work of Thomas Kuhn (1962) is probably most closely associated with this move, but the recognition that knowledge was produced socially perhaps also resulted from the practices of science in the 20th century that involved greater collaboration than it may have seemed to in prior times (although arguably this perception is mistaken – after all, even Newton noted that he had seen further by "standing on the shoulders of giants"). Once the social nature of science is acknowledged, this gives rise to questions about which social factors might influence or be relevant to knowledge production – either aiding it or hindering it. How do gender, race, class, and ability play a role in knowledge production (if at all)? Feminist epistemology begins with a focus on gender. As already noted, one of the more widespread features of societies is the way that power is unequally distributed, and gender is one of the matrices along which that unequal distribution is made.

The question of whether and how social factors such as gender might influence or be relevant to knowledge production is not merely an academic question, it has a range of significant consequences. Knowledge itself is a source of power. Knowledge enables well-grounded decisions and actions. Scientific knowledge is important for determining how best to address a variety of environmental, social, and health challenges that threaten wellbeing. The production of knowledge also goes hand in hand with the development of new technologies that have the power to shape our lives and communities. How we understand the world around us provides us with opportunities to both protect and change that world. Knowledge enables us to understand our own experiences and decide how we want to live our lives and interact with others. In addition, the demand for those trained in knowledge-producing professions, including those in science and engineering, is increasing and provides significant economic opportunities. Thus, the extent to which intersecting social factors, such as gender, race, class, and ability, influence who can participate in knowledge production, what we know, and how we know it can be incredibly important for ensuring that knowledge production is a benefit to be equitably shared, rather than benefiting only a few.

Whether and how social factors such as gender or race influence the production of knowledge is also important for avoiding harms. When we test whether a drug is safe and effective, we want to know whether it will be safe for *everyone* who might be likely to take it. When we observe non-human animals, we want to ensure that we are not merely projecting our own social roles and stereotypes onto what we observe. When we study the human brain, we need to be sure that the knowledge we are producing is not influenced by assumptions or stereotypes we may already have about differences between sexes, races, classes, or those with diverse cognitive abilities. In other words, we want to ensure that social stereotypes and biases do not inappropriately influence assumptions, experiments, or what we conclude from evidence. Finally, if social factors can influence who can fully participate in epistemic or scientific communities, this may create harms by limiting the opportunities of particular groups, but also perhaps limiting the knowledge that is produced in ways that might be harmful to everyone.

In the next section, we examine the questions that will guide the rest of this book to explore the ways in which social factors may influence knowledge production in more detail. These questions are aimed at trying to understand how social factors such as gender can influence the production of knowledge in different ways that can limit the benefits of knowledge and produce harms, so that we may consider how to best address this.

We explore each of the six questions raised in the following chapters, paying attention to how feminist epistemologists and philosophers of science have addressed them, contrasting their approaches with more traditional views, and noting where they share similarities with other critics of traditional views. These questions are of general interest for any theory of knowledge, but often they were not consciously addressed. Sometimes the answers were presupposed and unexamined and feminist attention to these questions has been one factor in shifting how we currently understand these areas of philosophy.

Who knows?

As an example of how feminists have shifted thinking, consider the first of our questions: Who knows? Various approaches to investigating belief, knowledge, and understanding differ from each other in many ways, but they have primarily focused on what is known rather than who knows it. This approach meant that assumptions about who knowers are and what characteristics and capabilities they should have were left unexamined. Feminist epistemologists pointed this out and questioned traditional assumptions about knowers.

Traditional epistemology often makes assumptions about what cognitive capabilities are necessary for knowledge, for instance. Typically, reason and sense experience in some appropriate combination are thought to be needed to produce knowledge. Although reason is a feature that is sometimes attributed to all humans, historically many philosophers explicitly denied that women (and children) were fully capable of the appropriate use of reason, a move that seems contrary to the assumption of reason as a universal human trait. Aristotle, for example, thought that women were not equal to men in several ways, including in their ability to exercise reason. He notes that their reason "lacks authority". Kant has also been singled out by feminist philosophers as holding views that denigrated women's ability to reason. Human knowledge is understood by such philosophers as that of which only those who are fully human are capable and there is debate over who among all humans qualifies as fully human. Women have failed to meet this standard in the judgment of many philosophers.

In any case, if we set aside the question of which capabilities are needed and who has them, there is another aspect to traditional approaches to knowledge that is worth noting. Traditionally, the characteristics of knowers that are deemed relevant are cognitive capabilities. As noted in the primatology case we started with, the social location of the knower – the historical, political, and cultural context in which knowers find themselves including their gender, race, ethnicity, sexuality, or ability – is not thought to be relevant. As Lorraine Code puts it, "philosophers commonly treat 'the knower' as a featureless abstraction" (Code 1991:1). She notes the traditional formula for epistemological analysis has been "S knows that p if and only if certain conditions hold" and the focus has been on identifying what conditions are necessary and sufficient for achieving knowledge about p (a particular proposition). Epistemologists tended not to ask: *who* is S? That question was not taken to be relevant, as evidenced by the practice of reducing a knower to a mere variable (S), stripped of any social features. What is also apparent in this formula is that the knower is conceived of as an individual. Knowers are isolated individuals who simply know in virtue of their own cognitive capacities. The individual knower is thus abstracted from all the conditions and relations in which actual human beings find themselves. Actual humans are embodied, existing at a particular place and time, and related to other individuals in a variety of ways, while an abstract knower is not.

While philosophical writings may have discussed and debated who has the cognitive capabilities required of knowers, they did not discuss their social location, nor consider the role that relationships with others play, in producing or attaining knowledge. This lack of scrutiny of the relevance of these factors is one aspect of traditional epistemology that feminist approaches challenge and correct. In Chapter 2, we will explore the features of the traditional approaches that either denigrate women as knowers or presume the irrelevance of gender and discuss how feminists have argued that they are both problematic in different ways. We will follow the feminist argument that who knows matters (as does who does not know).

What do we have knowledge of?

Chapter 3 focuses on the question of what it is that we have knowledge of. Returning to the formula "S knows that p", p is traditionally understood as a proposition that must be justified and true to count as knowledge. A long-dominant view in epistemology is that when the conditions for knowledge are met, what we have knowledge of are propositions that describe some true states of affairs in the world that are independent of humans or human minds. On this view, knowledge is achieved when the subjective mind grasps something that is true about a mind-independent (objective) reality. This account of knowing assumes the possibility of a one-to-one correspondence between what we believe and some state of the world, as well as a singular conception of what it is to "get it right".

Broadly, feminists agree that what we know is the world, but what that means can be understood in different ways. Knowledge projects are typically focused on a particular subject matter. This is true in two ways. First, what is known is at least to some degree dependent on what is deemed to be of interest. Inquirers focus on particular objects of inquiry based on those interests. Inquiry is guided by the research questions that are asked and those questions are reflections of the interests of the inquirers. In the primatology example, female primatologists who entered the field in the 1970s and 80s were interested in the activities of female primates and the role of females in evolution – partly because these topics had been neglected in past accounts. This may be one reasons why they produced different observations that questioned prevailing assumptions. That is, they had different interests because of a variety of historical and social factors. Second, in considering a particular object of inquiry, only some aspects of that object may be thought to be worth exploring. Indeed, the activities of female primates that were not related to reproduction had also been previously neglected because they were not viewed as scientifically interesting from an evolutionary perspective. Both of these limitations on what we study in the world have been characterized by a number of contemporary philosophers in the following way: we are interested in truths that are thought to be significant or interesting – not necessarily in all truths.

Another well-known example illustrates this point. Until relatively recently research on heart disease had been focused on the disease as it occurred in men. This is in part because the presence of the disease in men seemed a pressing problem since it was the leading cause of death in men in industrialized nations. However, it turns out that it is also the leading cause of death among women, so why didn't the disease get the same attention there? At least one reason is because of the disproportionate economic impact that the death of working men in the prime of their lives was having. Women were not present in the workforce in the same numbers and so were not as visible. It also turns out that the course of heart disease and symptoms in women are often quite different to those experienced by men. With the focus already on men and heart disease understood through its presence in men as a paradigm, heart disease in women was often missed or misdiagnosed. This is a clear result of interests – in this case the interests of the dominant group – shaping the course of research. While it appeared that the object of inquiry was heart disease, in fact, it was heart disease in men. While research on women is more frequently conducted now, the lingering effects of the historical focus on men are still felt. A recent survey of the public perception of leading causes of death among women shows that the majority of the public and doctors are still unaware that heart disease is the leading cause of death in women (Bairey Merz et al. 2017).

Recognizing the importance of interests in guiding knowledge demonstrates the interactive nature of knowledge. Knowers influence the process of knowing and what is known. Combined with the understanding of knowers as socially located (as explored in Chapter 2), it suggests that variations in context are likely to produce different sorts of knowledge. While this result is trivial in some cases – e.g., because of my experiences traveling in Mexico I am likely to know different things than you if you have never been to Mexico – it is more problematic in the case of scientific knowledge and other knowledge that claims to be objective or universal. For example, it might be alarming if the social context of the knower was relevant to knowledge about the relationship between acceleration, force, and mass of objects. If the social context of a subject always influences what we know, this seems to imply a subjectivity to knowledge that would seem to threaten scientific objectivity. It may suggest an unacceptable relativism – or the view that what is true or counts as knowledge depends entirely on the social context or what an individual knower believes.

While most feminist epistemology and philosophy of science seeks to retain some notion of objectivity, the rejection of the abstract decontextualized knower and the acknowledgment that the social location of the knower matters is in tension with conceptions of objectivity for which traditional epistemology and philosophy of science aim. To put it more clearly, knowledge is knowledge of the world as it matters to us. In contrast, traditional understandings of knowledge seem to presume that there is some way to achieve what Donna Haraway (1988) calls the "god trick" – the trick of knowing the world from a God's eye point of view or as it would be without us. Thomas Nagel (1986) describes this as the view from nowhere. But to say that the world is known from some social location need not mean that it is not known as it really is – it is just not known completely or in totality – but this is hardly a controversial claim.

The importance of context for feminist conceptions of knowers and the related idea that all knowledge is situated and that the social location of the knower matters is a challenge to the traditional conception of objectivity and many critiques of feminist approaches center around this concern. The worry is that once the understanding of knowledge as a view from nowhere is abandoned, competing claims to knowledge will arise with no legitimate means of adjudicating among them or distinguishing knowledge from belief. Most, although not all, versions of feminist epistemology and philosophy of science dispute the traditional conception of objectivity and offer alternatives. We investigate these in Chapters 3 and 4.

As we explore feminist responses, a variety of other related issues emerge. The traditional picture as described in the previous paragraphs is one that presupposes a primary aim for knowledge production – that is, to accurately represent the world and the various connections among its parts. But the production of knowledge may have a multiplicity of aims. We want to know not only in order to accurately represent the world, but in order to affect change in the world. This is a pragmatic goal. Individuals also pursue knowledge for personal reasons – in order to advance their careers or for fame and prestige. Over the last several decades it has become more common to refer to the *aims* of science (or knowledge more generally) rather than the aim of science. Feminist philosophers of science have often embraced this pluralism of aims.

Additionally, the acknowledgment that the interests of knowers affect both the knowledge projects pursued and how what is investigated is understood evokes the role of values in knowledge production. To say that we study what matters to us is equivalent to saying that we study what we value. If this is correct, then this indicates at least one way that values are relevant to knowledge. Values matter in a variety of other ways as well. They may affect decisions about when to act on what we believe, including the decisions we make about how much evidence is required to support such action. There are many other ways that values enter into research and there is currently a wide body of literature on the role of values in science, some of which is the work of feminist philosophers. Again, this topic reflects the contrast between what had been a traditional view of good science as conforming to a value-free ideal and the view held by many feminists that feminist epistemology and philosophy of science should be informed by and reflect the values associated with feminist political goals.

How do we know?

How do we know is perhaps the most traditional of epistemological questions. This question concerns issues about the nature of evidence, that is how much and what kind of evidence is required for knowledge, as well as the nature of of confirmation, explanation, and prediction. Additionally, the distinction between knowledge and other cognitive attitudes (belief, desire, doubt, wishing) is central to epistemology. Traditional answers given to such questions have often involved the explication of the right method through which knowledge should be produced or the rational reconstruction of the logic of scientific inquiry.

The view that knowledge is dependent on having the right method is related to concerns about objectivity that are discussed in Chapter 3. Using the right methods can be one way to eliminate idiosyncratic beliefs or values that might be held by individual knowers. However, once we acknowledge that knowers are socially situated and that knowledge production is itself a social process, it is less clear how method can play this role. As feminist epistemologists and philosophers of science have urged the consideration of how gender norms and relations affect knowledge production, they have also explored our understanding of scientific method and evidence in a variety of ways.

As we discuss in Chapters 2 and 3, who knows and what they investigate makes a difference to the knowledge that is produced. Yet this can also make a difference to *how* things are studied, either because of the choice of objects of inquiry or in the methodologies used to study them. To understand why who knows and what is studied make a difference to how we know, it helps to recognize that talk of "*the* scientific method" is misleading. It suggests that there is one right way to go about producing knowledge when, in fact, knowledge production involves many different kinds of activities. In Chapter 4, we list nine different phases of research, each of which involves making decisions that may be influenced by historical, political, social, and cultural factors such as values. We have already noted that choices made about what subject matter to investigate and what research questions to ask affect knowledge production. These, in part, shape what observations are taken to be relevant (what counts as evidence). But researchers and other knowers also make choices about how to go about collecting evidence, how it should be interpreted, and when we have enough evidence to draw conclusions. These are unavoidable decisions that shape the research process and ultimately shape the knowledge produced. Given the social nature of knowledge such choices are subject to social influence. Feminist epistemologists and philosophers of science are among the philosophers who have noted these influences and been concerned about how to give an account of the objectivity of knowledge that takes into account social, political, and cultural influences while at the same time retaining the positive aspects of knowledge production that the ideal of objectivity seeks to capture. In Chapter 4, we explore a number of feminist approaches to how we know objectively, specifically those of Helen Longino, Sandra Harding, and Alison Wylie.

Longino's approach focuses on producing objective knowledge through providing guidelines for the social milieu in which knowledge is produced. She seeks to answer the question of how knowing communities should be organized and what norms they should adopt to produce objective knowledge. Harding and Wylie utilize feminist standpoint theory, an approach that offers researchers guidance on how to produce knowledge which is both objective and able to serve egalitarian aims. Each of these offers a way to think about how the presence of values in knowledge production can be compatible with the objectivity of knowledge.

Feminists in the pragmatist tradition have also addressed the question of values and objectivity. We consider Elizabeth Anderson's and Sharyn

Clough's arguments that empirical evidence has a bearing on values as well as beliefs. Consequently, determinations of objectivity for values do not differ materially from determinations of objectivity of beliefs.

What don't we know?

As should be clear from Chapters 2–4, traditional epistemology and philosophy of science have focused almost exclusively on the study of knowledge production (and of the production of scientific knowledge in particular). Philosophers have examined what it means to know, and in particular what counts as evidence or justification, as well as how much evidence we need to achieve knowledge. Yet the relatively new and emerging area of agnotology asks "what don't we know and why don't we know it?" Agnotology is the study of ignorance and examines the ways in which ignorance is produced and maintained. When we ask "what don't we know?" it may be tempting to answer, "everything that we don't yet have knowledge about!" Feminists and others working in agnotology, or what is also called "epistemologies of ignorance", have argued that ignorance is not merely a lack of knowledge (Tuana and Sullivan 2006; Proctor 2008). Rather, ignorance is often the result of complex social and institutional factors that can either intentionally or unintentionally obscure knowledge or prevent us from achieving certain kinds of knowledge. For example, Naomi Oreskes and Eric Conway (2011) have shown that private industry and think tanks have basically followed a playbook to undermine science, about things such as the health effects of smoking or climate change, and create doubt about the state of the evidence in order to stall regulation.

Feminists, critical race theorists, postcolonial theorists, and critical disability theorists were also all interested in the production and maintenance of ignorance, though perhaps for distinct reasons. While private industries may have a profit-motivation to generate and maintain ignorance about smoking or climate change, the consequence is to reinforce the status quo and prevent change. As mentioned, knowledge is power, and what we don't know can be a way of maintaining power or preventing social, political, or policy changes. Thus, several scholars have examined the ways in which institutional or social forces of sexism, racism, classism, colonialism, and ableism, have intersected to produce or maintain ignorance of knowledge that could be vital to understanding and addressing systems of oppression and systemic inequalities. Chapter 5 examines the ways in which these social influences operate to constrain and hinder what we know and how we know. This chapter also highlights the work in a relatively new area of epistemology and philosophy of science that feminists and others have brought into focus.

Why does it matter?

The questions and concepts that have traditionally constituted the focus of epistemology and philosophy of science are important precisely because we believe knowledge to have value. We want to understand what it means to know, who knows, what we know, and how we know so that we can identify what successful knowledge production involves and how we can improve it if needed. While much 20th century epistemology and philosophy of science was focused on analyzing the concept of knowledge and assumed that science was an ideal paradigm for successful knowledge production, we can also learn from past mistakes and successes to develop recommendations for how to know better. Feminist epistemology and philosophy of science have recognized and promoted this normative project. They have often focused on cases in the history of science to develop recommendations about how to achieve evidence, when to accept certain beliefs as fact, when to posit causal connections, what makes for good explanations, how to avoid bias, and how to structure scientific communities. Feminist contributions to these fields aim to enable us to better understand our epistemic aims and how best to achieve them.

Feminists, however, have also viewed these epistemic aims as deeply intertwined with social and political aims. Thus, while feminists agree that epistemology and philosophy of science are important for all the reasons above, they also believe that addressing those issues matters for addressing inequalities and achieving justice. Chapter 6 examines feminist work that explores the intersections that epistemology and philosophy of science have with ethics and political philosophy. We focus on the kinds of harms that can result from our knowledge-producing practices, giving special attention to the kinds of epistemic harms that can result. For several feminist scholars, improving our knowledge practices is deeply connected with addressing inequalities in access to knowledge-producing resources, identifying the ways in which stereotypes contribute to bias, addressing the ways in which systems of oppression limit what we know, and ensuring that the benefits of science and technology are more equitably distributed. Feminist scholars have worked to ensure that knowledge-producing practices achieve their liberatory potential, as opposed to just producing some true beliefs that may benefit a few. In addition, there is a growing number of scholars who have examined the ways in which limiting or hindering knowledge can produce a distinct kind of injustice: an epistemic injustice where certain individuals or groups are denied resources they need to make sense of their own experiences, or are prevented from having those experiences taken seriously. Chapter 6 examines how that can occur. Ultimately science and other knowledge-producing practices have the potential to either reinforce or challenge systems of oppression. Feminist scholars have examined how the liberatory potential of knowledge can be achieved.

How can we know better?

As we will discuss in Chapter 7, the answer to the question how we can "know better" may depend on what our conception of "better" is. This concluding chapter elaborates on the lessons of Chapters 2–6 for promoting epistemic practices that lead not only to knowledge that is epistemically sound or true, but knowledge that can also help achieve social and political ends and avoid epistemic harms. While it is surely an injustice to limit access to knowledge production, many of the accounts that we examine also make the case that knowledge production is itself harmed when particular groups are systematically excluded or limited in their participation. There are competing explanations for why this might be so and alternative accounts of how to best to organize research communities. We will conclude with some of the ways in which knowledge practices might be restructured or refocused to ensure that knowledge is not only reliable, but benefits all and provides the resources for identifying and addressing social inequalities.

Discussion questions

1.   How are the authors understanding "feminism" and how does this compare to your own understanding of this term?
2.   Why might one think that feminism is irrelevant to questions about knowledge, and especially scientific knowledge?
3.   What changes took place in philosophy of science during the 20th century that supported the possibility of feminist philosophy of science?
4.   What are some of the reasons that led feminists to question traditional assumptions in epistemology and philosophy of science?

References

Bairey Merz, C. N., C. J. Pepine, M. N. Walsh, J. L. Fleg, P. G. Camici, W. M. Chilian, and N. Wenger (2017). Ischemia and nonobstructive coronary artery disease (INOCA) developing evidence-based therapies and research agenda for the next decade. *Circulation*, 135 (11), pp. 1075–1092.

Code, L. (1991). *What can she know? Feminist theory and the construction of knowledge*. Ithaca, NY: Cornell University Press.

Haraway, D. (1988). Situated knowledges: The science question in feminism and the privilege of partial perspective. *Feminist Studies*, 4 (3), pp. 575–599.

Hrdy, S. B. (1986). Empathy, polyandry, and the myth of the coy female. In: R. Bleier (ed.), *Feminist approaches to science*. New York: Pergamon, pp. 119–146.

Kuhn, T. S. (1962). *The structure of scientific revolutions* (Vol. 111). Chicago: University of Chicago Press.

Nagel, T. (1986). *The view from nowhere.* Oxford: Oxford University Press.

Oreskes, N. and E. M. Conway. (2011). *Merchants of doubt: How a handful of scientists obscured the truth on issues from tobacco smoke to global warming.* Bloomsbury Publishing USA.

Proctor, R. N. (2008). Agnotology: A missing term to describe the cultural production of ignorance (and its study). In: R. N. Proctor and L. Schiebinger (eds.), *Agnotology: The making and unmaking of ignorance.* Stanford: Stanford University Press, pp. 1–33.

Tuana, N., and Sullivan, S. (2006). Introduction: Feminist epistemologies of ignorance. *Hypatia: A journal of feminist philosophy,* 21 (3), pp. i–iii.

2

WHO KNOWS?

Overview

This chapter explores two ways in which knowledge production is social. First, it considers the importance of the social location of the knower. Feminist epistemologists and philosophers of science have argued that all knowers are situated in specific social locations – locations determined by their gender, sexuality, race, class, and (dis)ability – and that the social location of the knower matters for knowledge production. Second, feminist philosophers of science, have identified a variety of ways in which knowledge is dependent on others through testimony, trust, and the need to rely on expertise. Scientific research is typically engaged in by research communities which are themselves situated in broader historical, social, cultural, and political contexts. Various approaches in feminist philosophy of science have examined how the social nature of knowledge affects knowledge production through considering the ways in which we are epistemically dependent on others and how that dependence is impacted by social norms and relationships.

Introduction

In 1990, Lynn Hankinson Nelson published a book that asks the question that is the title of this chapter – *Who Knows?* While feminist thought and feminist approaches had been around for many generations, it was only during the 1980s and 1990s that philosophical feminism began to explore the various ways in which knowledge might be affected by gender. Hankinson Nelson's title reflects one important aspect of how this move took place – through questioning previous assumptions about knowers. She was not alone

DOI: 10.4324/9781032693781-2

in asking such questions. Several other feminist works from this period share this focus. Sandra Harding's *Whose Science? Whose Knowledge?* and Lorraine Code's *What Can She Know?* both appeared the following year (1991). These contributions to feminist thought indicate an emerging awareness that philosophers had been focused on what is known rather than who knows it. If we think of the formula S knows that *p*, the question of how *p* – a belief, proposition, or sentence – can be known focused on what sort of evidence or procedures were required in order for *p* to count as knowledge. This focus hid several implicit assumptions about S. Philosophers traditionally had in mind a knower of a particular sort – isolated, individual, autonomous, and detached.

We mentioned the examples of Aristotle and Kant in Chapter 1, but there are many instances in the history of the philosophy of Ancient Greece, Medieval Europe, the Renaissance and Modern period where women are explicitly identified as somehow unfit to be knowers. However, such exclusions need not be explicit. Consider the knower in Descartes's *Meditations on First Philosophy*, for example. While Descartes does not claim that the knower must be male, and in fact respected and engaged with female knowers as evidenced by his extensive correspondence with Elisabeth of Bohemia, he treats knowers as generically human and he makes implicit assumptions about what that entails. The knower of Descartes's *Meditations* is an isolated, disembodied mind whose knowledge depends entirely on the use of reason and innate ideas, both presumed to be present in all human beings. It is through the proper application of reason – the right method – that knowledge is ultimately possible for any rational being. But this seemingly progressive Enlightenment conception of knowers turns out to be more problematic than it might at first appear.

Feminists critique this notion of abstract knowers in a variety of ways. We will focus on two in this chapter. First, there is the worry that the abstract knower is in fact not generic but reflects the characteristics of its inventors – men situated in dominant positions in society. Although the traits attributed to knowers are often attributed to all humans and so thought of as universal traits, philosophers had implicitly assumed that all humans were like themselves – men with the freedom and means to pursue their interests.[1] The role of the social and material world (social position, wealth, leisure time, access to education, and, of course, gender) did not appear to be relevant to those for whom these were not barriers to knowledge. Feminists pointed out that actual knowers are socially, culturally, economically, and physically located and so unlike this abstract knower. They noted that the social location of the knower affects how and what can be known and so any account of knowledge should acknowledge the social and material context of knowers. This critique will be examined in more detail in the next two sections, "Women's ways of knowing" and "Situated knowledge".

A second related critique elaborated that knowing does not occur in isolation but rather depends on others both in ordinary and scientific contexts. We depend on others to develop or acquire epistemic resources such as language and concepts that are necessary for knowledge. In some cases, we depend on the testimony of others to form beliefs or to provide justification for our beliefs. Expert testimony is also sometimes necessary to increase our understanding or improve our decision-making in cases where we may not have relevant expertise. We sometimes depend on others to either confirm or disconfirm our own perceptions when we may be uncertain of our own reliability. The production of scientific knowledge also cannot occur in isolation. Generating scientific knowledge about some phenomena – such as climate change – involves many areas of expertise that cannot all be mastered by one individual. In such cases, science requires division of labor and collaboration among scientists in order to produce knowledge. Scientists also depend on the work of other scientists in order to build on previous discoveries. Indeed, the scientific practice of peer review occurs because it is often necessary to receive critical feedback on assumptions, methodologies, interpretations of data, or reasoning. All of this suggests that we depend on others in multiple and complex ways in order to form beliefs and have justification for those beliefs. Thus, any account that focuses on isolated individuals as primary knowers does not fully recognize the social and relational nature of knowledge.

This second critique, that knowledge is social, is addressed in the last section of the chapter.

"Women's ways of knowing"?

While the philosophical understanding of an ideal knower might be intended to be generically human, many of the traits attributed to this knower are stereotypically identified as masculine. Knowers are described as rational, logical, unemotional, detached, and consequently objective in some sense of that term (see Chapters 3 and 4 for detailed discussions of objectivity).[2] In contrast, women are often stereotypically identified as emotional, compassionate, connected to others, and intuitive. These "feminine" traits are typically not seen as conducive to knowledge, but, on the contrary, are thought to be problematic. These stereotypes have long been used to justify the under-representation of women throughout knowledge production. As recently as 2005, Lawrence Summers (who was then President of Harvard University) suggested that it was possible that the low representation of women in science and engineering might be due to a difference in innate aptitude in mathematics, an idea that is reminiscent of such views.[3]

One way of responding to stereotypes is to show that women do indeed have the desirable traits of knowers and that the stereotypes are mischaracterizations. Another possible response is to suggest that these traits are not an accurate characterization of what knowers require to participate in the production of knowledge. An elaboration of this idea is the claim that so-called feminine traits support knowledge as well or better than stereotypically masculine traits or, perhaps, that they support a different, but equally important, kind of knowledge. Some feminist thinkers have embraced this last approach and consequently valorized "women's ways of knowing".

The idea that there might be ways in which women know that differ from the ways in which men know does not necessarily commit one to the view that there are innate cognitive differences between men and women. Differences in the way boys and girls are socialized and expected to meet gender roles can shape their understandings of their own abilities, their interests, and their career choices. Stereotypes shape thinking about how we know and who can know and, thus, are self-reinforcing. As Evelyn Fox Keller notes, "The identification between scientific thought and masculinity is so deeply embedded in the culture at large that children have little difficulty in internalizing that identification" (1983a: 189). Following this line of reasoning, Keller (1983b, 1985) offers an account of the Nobel Prize winning work of biologist Barbara McClintock as breaking from the masculinist dominance of science, informed by her ability to develop a "feeling for the organism" which Keller attributes, in part, to McClintock's experiences as a girl and woman.

Although Keller describes McClintock's work as transcending gender rather than exemplifying feminine thought, Sandra Harding worries that Keller's account reinforces gender stereotypes rather than transcends them. "Keller mistakenly identifies feminism with the exaltation of feminine identity projects, rather than with exactly that transcendence of gender" (Harding 1986: 122). In other words, Keller seems to implicitly reinforce the view that McClintock, in virtue of being a woman, used capacities typically identified as feminine, such as empathy, and that this is part of what explains the success of her approach. Harding's assessment that it is an error to counter a masculinist understanding of science with some form of contrasting feminine identity is echoed by much subsequent feminist epistemological assessment of knowers. Several additional feminist arguments have led to the rejection of the women's-ways-of-knowing approach.

First, if women were to have a distinct way of knowing merely in virtue of their biology, we would expect to see evidence of sex differences in cognitive ability or cognitive capacities. Although a number of studies purport to show differences in cognitive abilities between boys and girls, or men and women, much of this research appears to implicitly incorporate sexist or other problematic assumptions (see the example of the rod and frame test for visual spatial abilities in Chapter 4). Feminist critique of sex difference research in neuroscience has been particularly pointed, noting that much of this research incorporates questionable theoretical assumptions such as brain organizational theory (Bluhm 2021) or sex essentialism (Bentley 2021).[4]

Second, if one argues instead that there are gender/sex differences that are due to socialization rather than biology, it is problematic to attribute any set of characteristics to *all* women in virtue of socialization since to do so ignores important differences *among* women. Gender is not the only social parameter that might affect how and what knowers know. Class, race, ethnicity, sexual orientation, and (dis)ability affect and interact with each other and gender in a variety of ways. We turn to the question of how such features affect the identify of knowers, define their social location, and consequently may affect knowledge production in the next section.

Thus, most feminist philosophers have rejected the idea that there is a distinct "women's way of knowing" or that all women in virtue of their biology have distinct epistemic or cognitive capacities. Nonetheless, one need not reject the idea that the gender of knowers can still be a significant factor in knowledge production. How this idea has been advanced is explained below.

Situated knowledge

One of the key features of feminist epistemology is the claim that knowledge is situated. To claim that knowledge is situated is to claim that knowledge is *for* and *by* socially located knowers and so both the production of knowledge and the knowledge produced is always *local*. The idea that knowledge is situated involves not only a rejection of the abstract knower but also a rejection of knowledge as timeless and universal – true everywhere and always – a Platonic conception of knowledge.

Two lines of critique of traditional epistemology stem from the understanding of knowledge as situated. First, as we have discussed in the previous section, when understood as situated, the knower of traditional epistemology has characteristics more properly attributable to men in particular social roles. Because such knowers occupy socially dominant positions from which they both produce, use, and evaluate knowledge claims, the assumption that their characteristics are shared by all knowers had not been obvious and so had not been challenged.

Second, the generic knower of traditional epistemology is also described as an isolated individual and so the fact that knowers are embedded within particular social situations is not considered relevant. The social nature of knowledge is not acknowledged. In this section we focus on the first critique. We turn to elements of critique related to the social nature of knowledge in the next section.

The term "social location" is a metaphor and, as with all metaphors, it requires some unpacking. One way to think about it is through the idea of perspective. We all know that things look different from different spatial locations. A table may appear rectangular when looked at from above but appear to be a trapezoid when we stand off to one side. Perspective gives us some idea of what might be meant by situated knowledge, but it is not completely satisfying. The perspective of an observer can result from a choice the observer makes about where to stand, but choice is not as straightforward in the case of a social location. Social location is, at least in part, a matter of pre-existing social, cultural, and political frameworks. Where each individual fits into such frameworks is to some degree determined by factors that they may have little power to alter, such as how they are racialized, what gender role they are expected to play, what their socio-economic status is, and how they are (dis)abled. Social location involves an interplay among the individuals that make up society and social, political, physical, and cultural institutions and norms.

In addition, although perspectives may vary, there may be a general agreement on what the characteristics of the world are and so some perspectives may be thought to be better than others. Indeed, this is the case in the example of the table. We are likely to conclude that the tabletop has the shape of a rectangle but appears to be a trapezoid from some non-optimal perspective. The observer who views the table directly from above and sees the table as a rectangle is deemed to have the appropriate point of view. This "view from above" has a counterpart in the generic knower of philosophical accounts who is thought to be in the right position to have knowledge. While this is not a literal view from above, it is the result of performing what Donna Haraway calls the "god trick" (Haraway 1991). The ideal viewpoint of the generic knower is how God might know the world – the view from everywhere and nowhere – a view which is not specific to any particular social location. It is often thought that achieving such a view is possible through adopting the right attitudes and using the right methods – methods that eliminate the particularities of social location and so achieve a kind of objectivity. In this way, social location is made irrelevant through the methods used to produce knowledge. We will discuss this idea in Chapter 4 when we tackle the question of how we know. Feminist epistemology challenges this understanding of knowledge by reminding us that actual knowers are always situated somewhere – most relevantly they are situated in terms of the gender, race, class, (dis)ability, and ethnicity categories of the society in which they live.

For these reasons, social locations are not the same as perspectives. If we further explore the spatial metaphor, we can think of social locations as given through matrices of various social, cultural, and political factors. The intersecting points of these matrices are social locations that might be thought of in much the way that longitude and latitude identify geographic locations. But for social location there are more than two parameters creating a myriad of different social locations. While sex/gender might be considered the most salient since our focus is feminist concerns, all women are further socially located economically, in terms of ability, sexual orientations, race, ethnicity, and so on. Within any society, power is differentially distributed across these identity categories in ways that affect access to the means of knowledge production.

Kimberlé Crenshaw introduced the term "intersectionality" to capture these complexities of social location and to emphasize how consideration of an individual's identity relative to merely one of these aspects will often fail to capture important features that result in the oppression or marginalization of individuals (Crenshaw 1991).[5] Intersectionality can be a useful analytical tool for thinking through how the situatedness of knowers can be relevant to knowledge production.

The marginalization of knowers can occur in a variety of ways: by directly denying their cognitive capacity as knowers, by failing to recognize them because they occupy social positions assumed to be inappropriate for knowers, by denying them access to the means of knowledge production (e. g., restrictions on education, structural obstacles to educational resources, or making the ability to acquire knowledge physically or intellectually inaccessible to those who are differently-abled), or by failing to recognize the relevance of their first-hand experiences as evidence (e.g., as occurs when testimony is ignored, dismissed, interpreted in ways that make them implausible, or silenced). While women may be marginalized in these various ways (discussed in more detail in Chapters 5 and 6), they are also marginalized differently depending on the specifics of their social locations. Intersectionality urges sensitivity to these differences and the problematic results of ignoring them. The origins of intersectionality in Black feminist thought include a Black feminist critique that White feminists had reproduced an error they had attributed to traditional philosophy. White feminists had assumed that knowers were like themselves and so, that the issues that White women faced were the same issues for all women. White feminism had consequently failed to address the oppression of Black women specifically but, more generally, that of women of color and all women who are otherwise marginalized.[6]

This brief account of situated knowledge should make it clear that to say that knowledge is situated raises a number of issues, not just for traditional epistemology but also within feminist epistemology. In particular, feminist epistemology has been accused of a kind of subjectivism or pernicious relativism since the idea that knowledge is situated suggests that knowledge is always relative to the beliefs of some specific knower. This misinterpretation has led some to claim that feminist epistemology is incongruous (Haack 2003: 8) in that what it describes is not really knowledge given this relativity. The traditional conception of knowledge treats knowledge claims as independent of knowers while feminist epistemology does not. However, to say that the social location of the knower is relevant to knowledge production does not thereby commit feminist epistemologists to subjectivist relativism – the idea that what is true is relative to the knower – but rather to taking seriously the relevance of social location in justifying and assessing knowledge claims.

Social location may be relevant in ways that do not imply subjectivist relativism. Researchers may be important contributors to knowledge in that their first-hand experiences may count as knowledge or provide important evidence for knowledge claims. For example, Alison Wylie discusses how grass roots researchers in the 1980s identified a "chilly climate" as one of the factors responsible for women leaving academia (Wylie 2011). Wylie notes that the project was partially motivated by the researchers' own first-hand experience of discomfort within the academy. The Study on the Status of Women in Science at MIT, which appeared in 1999, documented similar issues specifically in the sciences. In addition to their own experience, researchers conducted interviews in which they repeatedly heard stories of difficulty in accessing lab space, assumptions made about their cognitive abilities, tracking of women into service roles in the institution, and similar problems. Here, both the first-hand experiences of the researchers and those they studied serve as evidence of a phenomenon that came to be labeled "chilly climate" in the academy and the sciences. Wylie notes that these research projects were "local and internal" – and so situated. However, she also points out that this first-hand experiential evidence is not the only evidence supporting the conclusions of chilly-climate research. Research on bias from a variety of sources – e.g., psycholinguistics, sociology, cognitive psychology – together with first-hand experience makes a compelling argument that the problematic environment in the academy particularly in the sciences, bears some responsibility for the low numbers of women in these disciplines. In this particular case, we see that while this knowledge depends on knowers – some of the evidence is evidence of knowers' experience – it is also consistent with what else is known. It is knowledge for, by, and about women but not in the sense that it is only true for them.

Social location of the knower may also be relevant in that the experiences of researchers in a particular location may give reason to question traditional assumptions in their area of research. This may change what they think counts as evidence. Elizabeth Anderson illustrates this through a detailed account of research on divorce conducted by Abigail Stewart and her research group (Anderson 2004). Historically, studies on divorce had focused on harms to children that would result from the rupture of a household divided along stereotypical gender norms, where men would provide financially and serve as a "masculine" role model for boys, and women would engage in "complementary" domestic labor and serve as a "feminine" role model for girls. Under this framework, it would be difficult to *not* identify harms to children, since harms are defined in relation to those norms. Stewart's team shaped their research through feminist critique of the traditional family structure that led to their questioning whether the standard understanding of divorce as a one-time trauma and always as a negative event in the lives of those involved was the best way to think about it. They chose instead to view divorce as a life transition and a reconfiguration of family. As a result, they were open to considering both negative and positive effects of divorce. Anderson argues that the framing of divorce as trauma had directed previous research to focus only on evidence of negative effects. Consequently, previous research failed to consider evidence of positive effects, such as increased independence and autonomy for mothers, freedom from those gender norms, or decreases in verbal or physical abuse in households. Stewart's team's reframing of divorce in an alternative way allowed for consideration of evidence of both types, thereby also allowing assessment of feminist values in divorce (such as autonomy, freedom, or resistance to gendered norms and practices). The situation of researchers as feminists is what made them sensitive to the alternative framing and moved them to work with this different understanding of divorce.

Anderson's analysis also reveals that both approaches to researching the effects of divorce – understanding divorce as trauma and understanding it as a change in family configuration – incorporate value judgments. Divorce is a "thick" concept – that is, a concept that has both empirical and value aspects, as do many social concepts. Although this research depends on the social location of the knowers (in this case as feminists), it is not thereby relative to the knower, but, in fact Anderson argues, Stewart's account is more empirically adequate than previous research. However, because of the value-laden nature of the term "divorce", it might be deemed relativist in that it is relative to the particular value assumptions made. But both previous research on divorce and Stewart's research incorporate values. Neither approach is value-free and in each case the values reflect the interests of the researchers.

The social location of the knower may also affect the topics investigated, the questions asked, and what would count as a successful answer to that question – in other words, it may affect the aims of knowledge. As we discussed in Chapter 1, feminism is inherently political and so feminist epistemology includes liberatory aims in addition to the aims of traditional epistemology. For example, feminist approaches in medicine will focus on women's issues in circumstances where they have been neglected. We have already mentioned the case of heart disease in Chapter 1. The failure to study heart disease in women resulted in unequal treatment and so the egalitarian goals of feminism play a role in motivating research to correct that neglect. Again, such a shift of topic and change in research question does not result in a subjectivist relativism, but a more complete response to worries about relativism will come in Chapters 3 and 4 when we discuss feminist accounts of objectivity.

While the thesis of situated knowledge underpins much feminist epistemology, it has come to be most closely associated with feminist standpoint approaches. Feminist standpoint approaches are characterized by two core ideas: (1) a thesis of situated knowledge and (2) a thesis of epistemic advantage or privilege,[7] also called "the inversion thesis" (Wylie 2003; Wylie 2012; Tanesini 2019; Wu 2022). The thesis of epistemic advantage is the idea that a marginalized social location may offer an epistemic advantage because those who are marginalized must negotiate their way as outsiders in a dominant culture. They need to both understand the workings of the dominant culture and their own subordinate location within it. The result is that their social location can afford them insight into their own oppression and an understanding of the role of power dynamics in society. The "inversion" is that those who are subordinated are better knowers (at least about some things) than those who are dominant.

Standpoint theory has its roots in Marxism, where the economic structure of society is the focus and social locations are understood primarily as class locations. Feminist theorists extend the understanding of marginalized social location to include gender and ultimately other categories in an intersectional analysis. It is important to remember that standpoint theorists do not claim that members of marginalized social locations have epistemic advantage automatically. Belonging to a marginalized social group does not mean that one will immediately have access to knowledge of the social structures and assumptions that prevent the dominant perspective from having interest in the topics, goals, and evidence that are relevant to the lives of those who are marginalized. Understanding and the consequent shifts in knowledge only come about when standpoint is "achieved" through dialogue, thinking through, and working with others. Standpoint requires "studying up" – that is, starting from the local and immediate experiences of those being researched and moving to more abstract concepts through which to theorize that experience. For example, de Melo-Martín and Intemann (2011) consider why the human papilloma virus (HPV) vaccines have done so little to reduce global cervical cancer morbidity and mortality, even though clinical trials found them to be highly efficacious in preventing HPV-infection (and thus cervical cancer which is caused by HPV). This happened because the vaccine was developed and tested primarily by North American and European researchers and, indeed, that is where the vaccine has had the most significant impacts. The vaccine as developed, however, is highly difficult and expensive to manufacture and distribute because it requires refrigeration and must be administered by a medical professional. As a result, the vaccine has been less used by the populations of women who are most at risk of cervical cancer morbidity and mortality: poor and rural women in the Global South. They are less able to use it because of their local conditions including lack of refrigeration, scarcity of medical professionals, and inability to travel long distances for the 2–3 shots required over six months. As de Melo-Martín and Intemann argue, this is likely because researchers were not fully aware of the local conditions where cervical cancer is the most prevalent. If researchers had engaged more directly with the populations that were most at risk of cervical cancer, they may have been more aware of these limitations and theorized parameters for vaccines that would have better met the needs of those populations (de Melo-Martín and Intemann 2011). By "studying up" from the experience and conditions of those groups disproportionately impacted, such vaccines might have been developed in more effective ways.

As in the HPV example, studying up can occur through participatory research in which the researcher interacts with those studied to develop concepts and approaches that are a better fit to their lives.[8] Studying up can also occur through consciousness-raising – dialogue among those who are oppressed in which they identify and theorize their oppression. This, for example, has occurred in groups of women who began to recognize patterns in their experiences (e.g., being ignored in discussions with male colleagues, being touched in ways that made them uncomfortable, or having to endure comments that were sexual or misogynistic). In virtue of discussing and seeing these patterns, women began to theorize conceptual tools to name and understand their experiences (e.g., sexual harassment or microaggressions). This will be addressed in more detail in Chapters 5–7.

While the feminist understanding of knowledge as always situated indicates that knowledge is also always social, feminist standpoint theory points to another way in which knowledge is social. A feminist standpoint epistemology treats knowledge as social in that it involves social (and political) interaction to develop a standpoint. In this sense, standpoint is a social epistemology. But the relationship between social epistemology and feminist epistemology requires further exploration.

Knowledge as social: Feminism and social epistemology

Social epistemology came into its own late in the 20th century, both inside and outside of feminism. It is marked by a recognition of the social nature of knowledge production and how knowing depends on others in a variety of ways, both in terms of the relationships among individuals and the configuration of knowledge communities. These were themes that appeared in early feminist epistemology but also in philosophy of science.

There are at least three ways that feminist epistemology and philosophy of science identifies knowledge as social. We will refer to the first as the *relational* aspect of social epistemology. As we have emphasized in this chapter, in most feminist epistemology knowers are not conceived of as the abstracted, decontextualized individuals of traditional epistemology, but are recognized as embedded within communities and, in a strong sense, unable to know except as part of the community. Lynn Hankinson Nelson puts it this way:

> What I know depends inextricably on what *we* know, for some we. My claims to know are subject to community criteria, public notions of what constitutes evidence, so that, in an important sense, I *can* know only what *we* know, for some we.
>
> *(Nelson 1990: 255)*

Once we think of knowledge in this way, it is clear that epistemology should be investigating the relations among knowers and how those relationships might affect knowledge production.

Second, there is an *evidential* aspect to social epistemology. Much feminist work was strongly influenced by mid-20th century history and philosophy of science, particularly the work of Thomas Kuhn but also the Strong Programme in Sociology of Knowledge. The Strong Programme called for attention to the role of social values and biases in both failed and successful scientific theories, thereby indicating that social factors, while sometimes detrimental to knowledge production, could also play a positive role (Longino 2002). The examination of episodes from the history of science indicates that what is considered to be relevant evidence often depends on features of the political, social, and cultural environment in which knowledge is produced.

A third way in which philosophy of science can be seen as social epistemology has to do with the organization of the research community, its institutions, and the broader political, social, and cultural context in which the research takes place. This last way is the *organizational* aspect of social epistemology. Social contexts, including power dynamics and social structures, can influence the ways in which epistemic communities are organized and social epistemologists have recognized that communities can be organized in ways that facilitate or hinder the production of knowledge.

These three ways in which knowledge production is social (relationally, evidentially, and organizationally) may also be interconnected. For example, relational interactions within epistemic communities that involve establishing social standards for what constitutes evidence, will also reflect evidentiary judgments that are informed by social values. Similarly, how epistemic communities are organized will have implications for the kinds of relational interactions between community members as well as which values are represented in evidentiary considerations.

Nonetheless, we distinguish the three ways in which knowledge can be social because they also illuminate the ways in which feminist epistemologists have argued that the social features of knowers (including gender) are epistemically relevant. First, if knowledge is relational and depends on interactions with others, gender can influence those relationships and interactions among knowers. Gender may influence who is recognized as part of an epistemic community or how much credibility or expertise they are thought to have. Systems of oppression, including sexism, can hinder the perceived credibility or intellectual authority of marginalized individuals within epistemic communities. Second, social values – including gendered norms and political values – can influence evidentiary criteria for evaluating and accepting claims, or standards for what constitutes evidence. For example, those who value protecting the status quo might be more inclined to give weight to claims or evidence that coheres with other widely accepted views, while those who value transformational change might be more inclined to value novelty (Longino 1995). Third, feminist epistemologists have been interested in the ways in which gender has influenced organization of epistemic communities. For instance, feminists have examined how the exclusion or marginalization of women in scientific institutions and communities hinders knowledge production. Similarly, they have examined how epistemic communities might be structured in more egalitarian ways to promote knowledge production. Again, while gender has been a primary focus, it is not possible to discuss gender without also noting the differential and intersectional effects of other social categories. Consequently, the organizational aspect of social epistemology is often dealt with by exploring how diversity in the composition of knowledge communities and institutions might be epistemically beneficial.

Conclusions

In this chapter, we have examined the ways in which feminists have critiqued how traditional epistemology has answered the question: Who knows? Feminists have argued that when one attends to who S is – who the knower is – the assumption that knowers are disembodied abstract individuals is revealed to be problematic. This is both because it ignores the ways in which knowing depends on interactions with others and because knowledge is socially situated in that gender and other features of the knower can sometimes make a difference to what is known and how it is known. The gender of knowers can matter – not because women and men have distinct ways of knowing – but because those occupying different social locations or categories are likely to have different experiences that produce differences in the access to certain kinds of evidence. Social features of knowers, including gender, can also give rise to differences in access to epistemic resources and credibility as knowers (an issue we will return to in Chapter 6). Finally, differences in the distribution of power give rise to differences in the interests of those who occupy different social locations. We explore this in the next chapter.

Discussion questions

1. Why have feminists thought that the idea of the solitary abstract knower is problematic? Do you agree?
2. What is at least one way in which the gender of knowers might make a difference to what is known or how it is known?
3. What are the three ways in which feminists have argued that the production of knowledge is social?

Additional suggested readings

Crasnow, S. (2018). Contemporary standpoint theory: Tensions, integrations, and extensions. In: P. Garavaso (ed.) *The Bloomsbury companion to analytic feminism*. New York: Bloomsbury Academic, pp. 188–211. Offers a history of standpoint theory and a discussion of who the knower is in standpoint theory.

Crasnow, S. (2021). Feminist science studies: Reasoning from cases. In: H. Grasswick and N. A. McHugh (eds), *Making the case: Feminist and critical race philosophers engage case studies*. Albany, NY: SUNY Press, pp. 73–98. A detailed examination of how Elizabeth Anderson uses Stewart et al.'s study of divorce and Elisabeth Lloyd uses the case of theories of the female orgasm to analyze how values of researchers affect the results of research both positively and negatively.

Harding, S. (ed.). (2004) *The feminist standpoint reader*. New York: Routledge. This collection provides some foundational readings on standpoint theory, particularly in Part I.

Hundleby, Catherine. (2021). Thinking outside-in: Feminist standpoint theory as epistemology, methodology, and philosophy of science. In: S. Crasnow and K. Intemann (eds) *The Routledge handbook of feminist philosophy of science*. New York: Routledge, pp. 89–103. A good, contemporary account of the epistemic contributions that standpoint can make by reaching outside academic feminism.

Toole, B. (2019) From standpoint epistemology to epistemic oppression. *Hypatia: A journal of feminist philosophy*, 14 (4), pp. 598–618. Discusses standpoint in relation to the traditional S knows that p formula and argues that standpoint theory sheds light on epistemic oppression, a topic related to discussions in Chapter 6.

Notes

1 It is important to acknowledge that there have been women doing philosophy throughout the history of philosophy. However, they have been fewer in number than men in part because historically fewer women had access to the kind of education that supports philosophizing. But even when women did make substantial

contributions to philosophy these have been undervalued and often forgotten in subsequent periods. A few examples from just the Modern period include the work of Mary Cavendish, Emilie du Chatelet, Anne Conway, and Elisabeth of Bohemia.

2 See also Genevieve Lloyd's *The Man of Reason* (1984) and Susan Bordo's *The Flight to Objectivity* (1987). Both discuss the close identification of reason with masculinity.

3 Summers put this idea forward as one of several hypotheses for the low representation of women in science and engineering. He noted that while there may not be a difference in means (average ability) there does appear to be a difference in variability with males outperforming females at the upper end (Transcript, *The Harvard Crimson*, February 18, 2005, https://www.thecrimson.com/article/2005/2/18/full-transcript-president-summers-remarks-at/).

4 Also in Fine 2010.

5 Crenshaw introduced the term in a legal context to clarify that anti-discrimination laws were written so that they focused on one aspect of identity discrimination and so were inadequate to address discrimination that resulted from multiple (intersecting) aspects of identity (1989, 1991). Collins and Bilge and Hancock have argued that the concept of intersectionality had a long history in Black feminism prior to Crenshaw's naming it (Collins and Bilge 2016; Hancock 2016).

6 Crenshaw (1991) also worries about differences among women of color and how ignoring such differences can result in a failure to recognize and address the needs of subgroups within that more general category.

7 We will use epistemic advantage rather than epistemic privilege since epistemic privilege has more readily been interpreted as being automatically attained with a subordinate social location. This is not the intended implication and "epistemic advantage" seems less likely to have that connotation.

8 Feminist standpoint theory raises additional questions about the objectivity of knowledge with its thesis of epistemic advantage and its specifically political stance. We will address these in Chapter 4.

References

Anderson, E. (2004). Uses of values judgments in science: A general argument with lessons from a case study of feminist research on divorce. *Hypatia: A Journal of Feminist Philosophy*, 19, pp. 1–24.

Bentley, V. (2021). Feminism and cognitive science. In: S. Crasnow and K. Intemann (eds.) *The Routledge handbook of feminist philosophy of science*. New York, NY: Routledge, pp. 328–339.

Bluhm, R. (2021). Neurosexism and our understanding of sex differences in the brain. In: S. Crasnow and K. Intemann (eds.) *The Routledge handbook of feminist philosophy of science*. New York, NY: Routledge, pp. 316–327.

Bordo, S. R. (1987). *The flight to objectivity: Essays on Cartesianism and culture*. Albany: State University of New York Press.

Code, L. (1991). *What can she know? Feminist theory and the construction of knowledge*. Ithaca, NY: Cornell University Press.

Collins, P. H. and S. Bilge. (2016). *Intersectionality*. Cambridge, MA: Polity Press.

Crenshaw, K. (1989). Demarginalizing the intersection of race and sex: a Black feminist critique of anti-discrimination doctrine, feminist theory, and anti-racist politics. *The University of Chicago Legal Forum*, 140, pp. 139–167.

Crenshaw, K. (1991). Mapping the margins: Intersectionality, identity politics, and violence against women of color. *Stanford Law Review*, 43 (6), pp. 241–1299. doi:10.2307/1229039.

de Melo-Martín, I., and Intemann, K. (2011). Feminist resources for biomedical research: Lessons from the HPV vaccines. *Hypatia: A Journal of Feminist Philosophy*, 26 (1), pp. 79–101.

Fine, C. (2010). *Delusions of Gender: How our minds, society, and neurosexism create difference*. New York, NY: W.W. Norton and Company.

Haack, S. (2003). Knowledge and propaganda: Reflections of an old feminist. In: C. L. Pinnick, N. Koertge, and R. F. Almeder (eds.), *Scrutinizing feminist epistemology: An examination of gender in science*. New Brunswick, NJ: Rutgers University Press, pp. 7–19.

Hancock, A. (2016). *Intersectionality: An intellectual history*. Oxford: Oxford University Press.

Haraway, D. (1991). *Simians, and women: The reinvention of nature*. New York: Routledge.

Harding, S. (1986). *The science question in feminism*. Ithaca, NY: Cornell University Press.

Harding, S. (1991). *Whose science? Whose knowledge? Thinking from women's lives*. Ithaca, NY: Cornell University Press.

Keller, E. F. (1983a). Gender and science. In: S. Harding and M. B. Hintikka (eds.), *Discovering Reality: Feminist perspectives on epistemology, metaphysics, methodology, and philosophy of science*. Dordrecht: Reidel Publishing Co., pp. 187–205.

Keller, E. F. (1983b). *A feeling for the organism*. New York: W. H. Freeman and Company.

Keller, E. F. (1985). *Reflections on gender and science*. New Haven: Yale University Press.

Longino, H. E. (1995). Gender, politics, and the theoretical virtues. *Synthese*, 104, pp. 383–397.

Longino, H. E. (2002). *The fate of knowledge*. Princeton, NJ: Princeton University Press.

Lloyd, G. (1984). *The man of reason: "male" and "female" in Western philosophy*. London: Methuen Publishing.

MIT, Committee on Women Faculty in the School of Science. (1999). A study of the status of women faculty in science at MIT. *The MIT Faculty Newsletter*, Vol. XI(4). http://web.mit.edu/fnl/women/women.html.

Nelson, Lynn Hankinson, (1990). *Who knows: From Quine to a feminist empiricism*. Philadelphia: Temple University Press.

Summers, L. (2005). Transcript. *The Harvard Crimson*, February 18, 2005. https://www.thecrimson.com/article/2005/2/18/full-transcript-president-summers-remarks-at/.

Tanesini, A. (2019). Standpoint then and now. In: M. Fricker, P. J. Graham, D. Henderson, and N. J. L. L. Pedersen (eds.), *The Routledge handbook of social epistemology*. New York: Routledge, pp. 335–343.

Wylie, A. (2003). Why standpoint matters. In: R. Figueroa, & S. Harding, (eds.) *Science and other cultures: Issues in philosophies of science and technology*. New York: Routledge, pp. 34–56.

Wylie, A. (2011). What knowers know well: Women, work, and the academy. In: H. Grasswick (ed.), *Feminist epistemology and the philosophy of science: Power and knowledge*, pp. 157–179.

Wylie, A. (2012). Feminist philosophy of science: Standpoint matters. *Proceedings and addresses of the American Philosophical Association* Vol. 86, No. 2, pp. 47–76.

Wu, J. (2022). Epistemic advantage on the margin: A network standpoint epistemology. *Philosophy and Phenomenological Research*. https://doi.org/10.1111/phpr.12895.

3

WHAT DO WE HAVE KNOWLEDGE OF?

Overview

Traditional epistemologists have argued that what we have knowledge of are propositions that in some sense, accurately or objectively, capture the world. While it is generally believed that knowledge should be objective, what this means is not always clear. This chapter identifies three broad senses of objectivity (agent, ontological, and methodological) and why feminist epistemology is often thought to be incompatible with each of these senses. We show that feminism challenges traditional conceptions of objectivity and we focus on how feminists have deepened our understanding of what it means to achieve objective knowledge about the world. Feminists have shown that what we know depends on the kinds of questions we ask, how research questions are framed, and how scientific phenomena are understood and represented. We identify two ways in which what we know is changed through feminist critique. The first is a change in scope. Women entering fields of scientific research explored topics that had previously been ignored. The second is through reconceptualizing what we know or inventing new concepts through which what we know can be understood.

Introduction

On April 20, 2021, Derek Chauvin, a former police officer with the Minneapolis Police Department, was convicted of the murder of George Floyd. The conviction was notable for many reasons, not least of which was the rarity of juries convicting police officers who are involved in the harm or death of those they interact with in the line of duty. Why were the jurors convinced by

DOI: 10.4324/9781032693781-3

the prosecution in this case when so many jurors in similar trials hadn't been convinced in the past? At least one reason may well have been that they were swayed by the consistency of the reports of witnesses to Floyd's death, but even more so by their own experience watching the video that the 17-year-old bystander, Darnella Frazier, had recorded. Referencing this video, prosecutor Steve Schleicher said in his closing statement, "It's exactly what you saw with your eyes. ... This wasn't policing. This was murder" (as reported in the *Washington Post*, 2021).

The prosecutor appealed to the jury to trust both what they had seen and the testimony of others arguing that this evidence supports conviction of the charge in a way that is not open to subjective interpretation or political views. The claim that the jurors could see for themselves what had happened relies on the idea that there are objective events in the world that can be established through evidence accessible by anyone. There are things that happen and there are ways to know that they happened. The conviction seems to support this view. The appeal of objectivity is that even in cases such as this, that are so emotionally, morally, and politically charged, biases that often shape beliefs can be overcome or made irrelevant by the right sort of evidence. It is indeed possible to know what happened. That it is possible to have knowledge of the world is a key presupposition of any non-skeptical epistemology, and importantly, of science.

But what is it that we are knowing? In the specific case of George Floyd's death, what is known is that an event had characteristics that meet the legal standard of murder. That is, there is something about the world that we are able to grasp in some way that allows us to correctly describe it. Such a claim has both an ontological and epistemological aspect to it. It is ontological in that it is about events in the world – what *really* happened – and epistemological in supposing there is a way to know what happened. The example illustrates what is at stake in the claim that some knowledge is objective. Without a belief that there is a fact of the matter about what happened and that there is some way to know it, it is hard to understand how a legitimate judgment could be made. In this chapter we explore what feminist critique of knowledge production offers in aid of understanding *what* we have knowledge of – the ontological aspect of objectivity.

We begin with a discussion of how the concept of objectivity has been understood. We identify three senses of objectivity, one of which is closely connected to thinking about *what* we know, whereas the other two are more relevant to how we know. We examine the ways in which feminism and feminist epistemology have been taken to be in tension with each of these senses. We then consider how feminist epistemology and philosophy of science challenge traditional understandings of objectivity (a project that will be continued in Chapter 4) and how, as a result, feminists have contributed to what we know. We argue that feminist research has altered what we know both through expanding the scope of inquiry and reconceptualizing the objects of inquiry. We close by returning to a discussion of the implications for objectivity.

What is "objectivity"?

Philosopher Ian Hacking describes "objectivity" as an "elevator word" – it valorizes some virtuous characteristics of knowledge practices or the representations they produce and in doing so "lifts" their status (Hacking 1999: 22). It is not at all easy to say what virtues "objectivity" is supposed to capture, however. Hacking argues that when we say that knowledge is objective, it is always in contrast with some other way it is *not* – e.g., it is not subjective, not biased, or not a conclusion drawn on whim. In this sense, the term does not have a positive meaning but rather indicates the absence of any of the vices that threaten successful knowledge production. He concludes that it would be better to do away with the term entirely and turn our attention to the identification and elimination of the offending vices.

However, many feminist philosophers (and others) disagree, thinking that objectivity does describe some meaningful property or properties of knowledge and indeed there are several reasons not to abandon it. First, it is in common usage and likely to remain so (John 2021). Second, looking at the various ways it is used, there is enough overlap among them that "objectivity" can be treated as a shorthand for a variety of virtues associated with knowledge. Some of the feminist approaches to objectivity that we review in the next chapter illustrate how this might be so (see the discussion of Wylie in Chapter 4). Still, the wide variety of virtues that the term "objectivity" is intended to indicate makes giving a general account of objectivity problematic.

Many scholars have offered accounts of the various meanings that "objectivity" has had (Lloyd 1995, Janack 2002, Douglas 2004, Daston and Galison 1992 and 2007). This literature makes clear that objectivity is a complex concept that can have different meanings that apply to different aspects of knowledge production. We highlight three broad ways of understanding objectivity here that will be relevant to this chapter and the next.[1]

These three understandings can be seen as connected to the first three questions of this book. We might ask who knows objectively? What is known objectively? and How can we know objectively?

First, when we ask *who* knows objectively, we are referencing the idea that objectivity is ascribed to knowledge that is free from the subjective or idiosyncratic biases of those who produced it. The jury's conclusion that George Floyd was murdered is objective in this sense insofar as the jurors who arrived at that conclusion did so based on the evidence and not, for example, due to biases any of them might have against law enforcement or the prosecutors in this case. We will refer to this first sense of objectivity as *agent objectivity*, as it focuses on the qualities of epistemic agents or communities who are producing knowledge. In traditional epistemology and philosophy of science, objective agents are often assumed to be detached, or able to distance themselves from any emotional or other personal attachment (such as their personal interests or values) they might have to the issue under investigation. This sense of objectivity has been particularly influential in philosophy of science, leading to the value-free ideal (VFI). Roughly speaking, the VFI proposes that (ideally) those involved in knowledge production (and specifically scientists or scientific communities), should not be influenced by moral, social, or political values but rather should only be guided by epistemic values – values thought to promote knowledge – such as empirical adequacy or truth. Scientists or scientific communities who allow non-epistemic values to influence scientific reasoning are likely to produce biased results and thus fail to meet the requirements for agent objectivity on this view.

A second sense of objectivity references what we know. We might think that knowledge is objective in the sense that it accurately captures some phenomenon in the world. The jury's conclusion that George Floyd was murdered is objective in this sense insofar as it accurately captures the events that happened and the qualities that Chauvin's actions had. The video of George Floyd's arrest was compelling evidence for Chauvin's conviction because it was causally connected to the actual event. If Chauvin had not treated Floyd as he did, there would have been no video of him doing so. The event is part of the cause of the video and, as such, it is the means through which the event is accurately represented and so evidence that the events occurred as the prosecutors claimed. We will refer to this as *ontological objectivity*, as it concerns the existence or status of that which we have knowledge of. To say that knowledge is objective in this sense is to say that we have gotten something right about the object under investigation or as Elisabeth Lloyd (1995) puts it, captured what is "REALLY real".

A third understanding of objectivity has to do with how we know objectively – that is, what procedures or methods should be used to produce knowledge. Some methods, practices, and procedures are thought to be reliable in achieving epistemic ends whereas others do not. Those methods, practices, and procedures that are reliable can be thought to be epistemically reliable in this sense. We will use the term *methodological objectivity* to refer to the use of methods, practices, or procedures that are epistemically reliable. The jury's conclusion in the George Floyd trial might be said to have been methodologically objective insofar as it was the result of procedures, rules for evidence, standards for reasoning, or forensic methodologies that were epistemically reliable in achieving accurate conclusions. Traditional philosophers of science have also assumed that science is methodologically objective, as evidenced by the epistemic success of science. Indeed, because scientific methods are often thought to be objective in this sense, they also are thought to lead to results that would be reproducible, regardless of who is employing them.

Thus "objective" is an adjective that can modify: 1) epistemic agents; 2) our accounts of the world; and 3) the methods that we use to gather evidence in support of those accounts. Of course, the three aspects of objectivity identified here are not wholly distinct from each other. Getting something right about the world, or achieving ontological objectivity, may depend upon having objective agents (those free of problematic biases or idiosyncratic beliefs) who employ objective methods correctly. Agent objectivity is required to use methods correctly and so needed for methodological objectivity. Both agent objectivity and methodological objectivity also aid ontological objectivity, in that they each contribute to getting our accounts of the world right. Another way of putting this is that the objectivity of agents and methods are considered to be a requirement for ontological objectivity. These connections reflect the entanglement of what we know and how we know – ontology and epistemology.

Despite this entanglement, we distinguish these aspects of objectivity analytically even though they are interconnected in practice. Ontological objectivity focuses on *what* we know. The claim that feminist approaches produce better knowledge might be understood, at least as a first pass, as a claim that they represent more accurately. The other senses (agent objectivity and methodological objectivity) are more closely tied to *how* we know. They have to do with the process of coming to know. Objectivity in these senses has to do with how the agent – the perceiver or the knower – is apprehending. Knowing requires adopting the right epistemic attitude, obtaining the right sort of evidence, and drawing warranted conclusions from that evidence. An attitude shaped by unwarranted and unexamined assumptions and beliefs may hinder understanding. Methods, practices, and procedures for producing knowledge aim at eliminating the influence of such biases.

With this sketch of the complex landscape of objectivity in hand, we will now consider why feminist epistemology is thought to conflict with the very notion of producing objective knowledge about the world.

Tensions between feminism and conceptions of objectivity

The very notion of feminist epistemology has been thought to be incompatible, or at least in tension, with traditional accounts of objectivity in all three senses discussed above. The claim that knowledge is situated, or shaped by political, social, and cultural factors has led some to think that feminist epistemology is incompatible with ontological objectivity. If knowledge depends in some way on social location, then it suggests that there is no way that things "really" are independent of that situated context. This raises concerns that feminist epistemology is committed to relativism – the idea that there is no objective knowledge able to be shared or agreed upon but rather what is known is relative to a specific social group or culture, a particular moral perspective, or even to a particular individual. Relativism is thought to be problematic – at least in some ways of understanding it – because if knowledge is always relative then it is unclear how we could defend some claims as more "reasonable", "justified", or "accurate" than others, except as relative to a particular context.

Feminist epistemology is also thought to be incompatible with agent objectivity. As noted in Chapter 1, there is a seeming conflict between feminism, which is explicitly a political stance, and the traditional view that knowledge should not be influenced or shaped by political commitments. The idea that objectivity requires knowledge production to be independent of political thought is one component of the VFI described in the previous section. Addressing this worry is not entirely straightforward since among the motivations for feminist critique are concerns about the existence of androcentric biases and sexist political commitments or values influencing traditional science. Addressing such biases through introducing alternative, feminist, commitments seems problematic and even paradoxical insofar as it appears to fight bias by calling for different biases. This troubling problem has been dubbed "the Bias Paradox": how can feminists advocate for the inclusion of feminist political commitments, if the problem with traditional science is that it has been influenced by sexist or racist political commitments? Several defenders of feminist epistemology have attempted to address this challenge. We will ultimately reconsider the Bias Paradox in Chapter 7, though resources feminists have for resolving or avoiding the paradox will become clearer when we consider alternative accounts of objectivity in Chapter 4. [2]

Finally, feminist epistemology has been thought to be incompatible with methodological objectivity, or the idea that there are methods that, if correctly employed, can achieve objective knowledge about the world. As we saw in Chapter 2, feminists have argued that *who knows* matters and that the social characteristics of knowers can influence what is known or how it is known. While feminists have tended to reject the idea that there is a "women's way of knowing", the idea that social demographics of knowers are epistemically relevant might be thought to be in tension with an idea that there are methods that are objective and reliable regardless of who uses them.

Despite these tensions, most feminists working on these issues have seen value in maintaining some understanding of science as objective. Consequently, a number of feminist epistemologists have developed accounts of objectivity that are distinct from traditional views of the three senses offered above and, in particular, do not equate objectivity with freedom from values. Their work in this area has been part of a larger debate about how to understand objectivity and whether the VFI is plausible and is one area where feminist work has influenced the work of non-feminist epistemologists and philosophers of science (and vice versa). Feminist critiques have contributed to and benefited from investigations of the role of moral, social, and political values in science and coinciding challenges to the VFI. Explorations of the role of values and the sorts of bias they might produce have included efforts to distinguish legitimate from illegitimate roles for moral, social, political, and cultural values. While much early feminist research involved exposing both explicit and implicit sexist bias that is damaging to knowledge production, more recently research has focused on ways that social and political values can be instrumental in producing better knowledge.

We will return to the question of values and science in more detail in Chapter 4 but, for the moment, we focus on the way values are implicated in debates about the ontological sense of objectivity with respect to what we know. The worry that ways of representing the world might be distorted by sexist or other politically charged values is a worry that accurate representation may not be possible once values come into play. As we noted at the outset of this chapter, "objective knowledge" has both ontological and epistemological dimensions. In the remainder of this chapter, we will focus on how feminist knowledge production has changed the ontology of various areas of research and in what sense the knowledge produced as a result of these changes can be understood to be ontologically objective. We will turn to the feminist contribution to thinking about *how* we know, as well as feminist accounts of agent and methodological objectivity, in Chapter 4.

Feminism and *what* we know

If (ontologically) objective knowledge is knowledge that captures "the way the world is", what exactly does that mean? "The world" is complex and multi-faceted. Knowledge seekers do not intend to know all parts of the world at once, nor even everything about some particular part of the world. Rather, they have something specific in mind that they want to find out about when they start their inquiry – although what that is may change as the investigation progresses. In Derrick Chauvin's trial, the job of the prosecutor wasn't to establish to the jury every single true fact about the events that happened at the intersection in Minneapolis the day of George Floyd's death (e.g., the clothes that every individual at the intersection was wearing). Rather, it was to establish those features that were relevant to the scope of their inquiry (i.e., whether George Floyd was murdered). Of course, various interests and values influence what the scope of inquiry is in a particular context. In Derrick Chauvin's trial, the interest was a legal one and so the focus of inquiry was directed to those features of the events that might have been relevant to whether the legal components of murder were met, or those things relevant to determining culpability. Yet, the fact that interests and values play a role in determining which features of the world we want to capture does not mean that there is no fact of the matter about whether George Floyd was murdered. Similarly, feminist scholars have argued that values and interests can influence what features of the world we are interested in while also acknowledging that some claims about the world or those features are more accurate than others.

The point we are making here resonates with a more general shift in philosophy of science that has taken place since the second half of the 20th century. As Philip Kitcher puts it, in pursuing knowledge researchers are not merely seeking truths, but rather *significant* truths (Kitcher 2001). This raises the following question: significant for whom? and significant for what purposes? Women brought a shift in topic and conceptualization to scientific disciplines resulting from differences in what they saw as significant. Such changes do not alter reality – that is to say, they do not change the world, but they do affect both what aspects of the world are studied and, as a result, what is known.

During the second half of the 20th century, as more women entered the academy and made inroads into the sciences, they brought with them new interests and approaches.[3] In some cases, the involvement of greater numbers of women resulted in changes to what was studied, as well as how research was carried out. For example, women in sociology found that some topics that interested them and were relevant to women's experiences seemed not to fit into the discipline's conception of appropriate areas of research. In economics, domestic work, the burden of which fell predominantly to women, did not figure in the study of labor. These gaps in knowledge were noted by women in these fields. What was known was not always what women in these fields wanted to know and what they wanted to know was often unexplored.

The engagement of women in research also had the effect of bringing to light unexamined assumptions about sex and gender that not only diverted attention away from some potential topics of research but also revealed ways in which such assumptions limited or distorted knowledge. We discussed an example of this in Chapter 1. In primatology, some women researchers approached their investigations through different interests than the men who had preceded them. The assumption that the behavior of dominant males was the central dynamic relevant to primate social structure resulted in very little focus on family units and, consequently, produced an incomplete understanding of primate behavior. As we discuss in detail in the section on problematic metaphors in this chapter, assumptions about the "maleness" of sperm and "femaleness" of the ovum resulted in an inaccurate lock-key model of reproduction in which the ovum was understood to be entirely passive.

Notice in these examples, it is not that researchers who were men failed to produce knowledge prior to the involvement of women researchers. The focus on male primates did indeed reveal something about dominance behavior, but it did not provide information about social relationships among female primates and their offspring – information that is relevant to a fuller understanding primate behavior. And, in the reproduction case, the lock-key model did provide understanding about how zygotes are created, it just obscured the complex interactions that play a role in that process. A shift in focus produced knowledge of these previously unexplored areas and so arguably improved the science. Researchers do not and cannot make observations of everything occurring around them, but they focus on what is considered of interest and those interests are partly a result of both explicit and implicit prior understandings of the area of study that shape the questions researchers pose. Some preliminary conceptions about the area of study circumscribe the scope of inquiry, including through shaping the research questions to be addressed. These reflections lead us to identify two ways that women working in knowledge producing fields brought about changes to what was known in those fields – through a focus on new topics previously not considered important by the discipline and through new ways of conceptualizing what was studied. In both instances, the ontology of research was altered.

As a way of clarifying this point, we introduce the term "objects of inquiry" to refer to those objects in the world that are the focus of inquiry in research (and understood in a particular way for the purposes of that research) as opposed to "the too plentiful and too various natural objects" that constitute the world (Daston and Galison 1992: 85). Objects in the world are those objects that exist, but objects of inquiry are those of interest in a particular research context. Once again, the point is that it is particular features of the multifaceted objects in the world that are the focus of any research inquiry.

We are not claiming that all women entering these fields researched different topics or conceptualized their areas of research differently than was traditional for their fields. Nor are we claiming that all women inquirers should be understood as feminists. As we have previously noted, not all women are feminists (nor are those who are feminists all feminists of the same stripe). What we offer here is a general characterization of some effects that women and feminists had and continue to have on changing science, and, more specifically, the accounts that feminist philosophers of science and epistemologists have offered to aid in understanding and characterizing those changes. Again, to be clear, when we say that the ontology of research is changed, we are not suggesting that reality itself is changed through these approaches, but rather that what is thought to matter about reality (what is considered relevant) is changed. Consequently, how reality is represented and what is known are altered.

As we saw in Chapter 2, among the most important insights of feminist epistemology's conception of knowers is that they are situated and that their location matters. That knowledge is situated, that it is always local, challenges the notion of a universal character of knowers (and knowledge). Differences in social location are relevant to *what* we know and so to what the objects of inquiry are for a variety of reasons. Knowers are situated in the sense that they have different social locations. They occupy different social "spaces" in the socio-cultural-political power structure and, because of this, they have differing levels of access to the means of knowledge production and the knowledge produced. Power differences result in differences in interests for those in subordinate positions as opposed to those in dominant positions. Consequently, differences in interests can result in some aspects of the world mattering more (or less) to those at the margins than to those who are closer to the center of power. This is one of the ways in which differences in social location can thus result in differences in what is known.

Consider for example, maps. Good maps will help us navigate the world correctly. But there may be different aims we might have that direct which features of the world we should represent or how to represent them. Two-dimensional world maps necessarily involve some distortion when representing a globe. The Mercator projection map (which is still standardly used) was developed in 1569 by Geradus Mercator as a navigational tool. Because the aim was navigation, it represents lines of latitude and longitude as parallel throughout so that navigators could have lines of constant compass bearings, making it easier for them to sail from Europe to North America. Yet, what the Mercator map distorts is the geographic size of some countries and continents relative to others. Quite notably, Africa is represented as smaller than North America, when in reality it is three times larger. The Peters projection map (designed by Arno Peters in 1973), by contrast does accurately represent the relative geographic size of countries and continents, but it does so by significantly distorting representation at the poles and along the equator. Which projection is "correct"? Both projections were useful in relation to their aims. The Mercator projection map was better for the aim of navigating. The Peters projection map, however, is better for understanding the relative size of countries and continents and it has been argued to be a "fairer" or more socially just map, precisely because it is not distorted in ways that make continents in the Global South less prominent or smaller than those in the Global North. Which map is "better" ultimately depends on the aims of map-users.

As this example illustrates, some of our aims in representing the world can be practical or political. Feminists also have political aims and so feminist science is an explicitly politicized science – a science which takes as among its aims the amelioration of oppression (see Chapter 1). Feminist goals reflect a set of liberatory values and so challenge the VFI and the sense of objectivity aligned with it. Understanding knowledge as situated means understanding that it is always partial, based in particular interests, and shaped by previous knowledge and interests. The interests and values of epistemic communities vary considerably across social locations and affect what is known in a variety of ways, as the map example illustrates.

First, because the world is complex and knowledge is pursued for specific aims, choices about what to study and, more specifically, what aspects of it to study, will affect the body of knowledge produced by any particular knowledge community, at any given time. Neither map in our example above represents all aspects of the earth. The lesson here is that knowledge is always partial.

Second, feminists stress that among the interests that shape what aspects of the world we are interested in are the specific aims for which knowledge is sought. Knowers have multiple aims and different features of the world are more or less useful for those aims. The map that supports the aim of successful navigation does not also support understanding the relative size of the continents. What features of the world are relevant at any particular stage of inquiry will depend on beliefs, desires, and needs, but beliefs, desires, and needs do not wholly determine what might be relevant for the successful achievement of aims. Some ways of thinking about the world will more successfully achieve a particular aim of inquiry than other conceptualizations or representations.

Third, what is known is dependent on the questions inquirers ask. How those questions are framed, why they are asked, and how scientific phenomena are understood and represented are all constrained by what is already believed to be the case (e.g., currently accepted theories), but also other background assumptions and beliefs some of which may be implicitly held and unexamined. Changes in the scope of inquiry will also change what questions are asked or what problems are deemed to need a solution. As we will see in the remainder of the chapter, women and feminists have altered what we know by challenging assumptions and beliefs in order to pursue the different research questions and achieve the different aims with their research.

As noted, we focus on two ways in which women inquirers had an impact on what was known. First, there was an expansion of fields of inquiry. Economics began to explore the role of domestic labor in the overall economy. Primatologists examined the family life of primates. Second, feminists reconceptualized the objects of inquiry. Sometimes they noted ways in which the investigation of specific topics had been shaped by sexist assumptions underlying past research. Explicit and implicit stereotypical understandings of women or the behavior of female animals were sometimes projected onto areas of research distorting its results. Research on hormones, for example, projected so-called masculine traits onto the "male" hormone testosterone and "feminine" traits onto estrogen. In addition, key concepts used in inquiry might fail to be adequate to respond to the research questions that women in these disciplines sought to address, as we saw in the previous chapter with Abigail Stewart's rethinking of divorce. We explore each of these in turn.

Objects of inquiry – expansion of scope

As a result of feminist critique and reflecting on the growing influence of women researchers in various scientific disciplines, we can identify some of ways in which "what" is known – the objects of inquiry – have changed. Many of these are discussed in the feminist philosophy of science literature (see especially Londa Schiebinger's *Has Feminism Changed Science?* 1999). Perhaps the clearest sense in which the scope of knowledge production has expanded is through new topics of interest. The most obvious example is simply the addition of women as worthy objects of inquiry. The failure to include women in the study of heart disease and the consequent lack of information about its symptoms and effects in women has been extensively discussed and is just one instance of a more general absence of research on women's health and women's bodies. Early feminist work pointed out these gaps in knowledge and sought to rectify them (Tuana 2006).

There are several other ways in which female animals or women might be missing as objects of inquiry. In economics, for example, the focus on labor outside of the home means that the domestic labor done primarily by women was not taken into account when developing economic theory or making practical decisions based on such theory. Feminist economists have pointed out the role of domestic labor (in fact coining the term "domestic labor" according to Barker and Kuiper 2021) in the overall functioning of the economy and contrived measures to indicate its monetary value, which cannot be directly measured by standard means since it is mostly unpaid. In addition, they have noted that it is an important consideration for the sub-field of development economics (the economics of the so-called "developing nations"). Early development approaches took the Northern industrial economies as a model – economies that emphasized labor in industrial settings. A consequence was that the role of women was de-emphasized. This proved problematic in that women were often involved in domestic and agricultural labor crucial to the economies of the non-industrialized nation-states.

Somewhat more subtle is the shift in focus in primatology where the previously understudied behavior of female primates resulted in modifying and improving the understanding of primate behavior. Similarly changes in paleoanthropology resulted from consideration of the role of female animals in human evolution resulting in accounts in which women played a more active role than had previously been ascribed to them. Schiebinger discusses the "women the gatherer" account developed in the 1970s by Sally Slocum, Nancy Tanner, and Adrienne Zihlman and notes that Zihlman specifically attributes the theory's development to the greater visibility of women generated by the women's movement of the era.[4]

These three examples indicate how failing to examine some parts of the world (in the cases above, female animals, women, or feminized areas of activity) can produce gaps in knowledge. To some extent, such gaps are inevitable (as will be discussed in Chapter 5). As we have already noted, the goal is not to know *everything* but rather to gain *significant* knowledge. However, in each of these cases, the failure to include women or the spaces in which they have traditionally operated clearly resulted in flawed understanding of the realm of inquiry. When women entered epistemic arenas they expanded interests thereby creating a broader and more inclusive body of knowledge.

Objects of inquiry – alternative conceptualizations

In addition to investigating topics that have not been adequately explored previously, feminist inquirers have also considered how the analytical tools through which the objects of inquiry are understood may shape or limit knowledge. We will explore three ways in which what is known is affected. First, the object of inquiry may be conceived through metaphors that affect both the way that we know and what we know. Second, the tools for inquiry and knowledge production may be inadequately applied because of assumptions about how the objects of inquiry should be conceptualized. Researchers may reconceptualize these objects making them better suited to the goals of research. We call this ameliorating concepts. Third, there may be no clear concept that captures what researchers want to understand, investigate, and study. This calls for the invention of a new concept. In all three cases, how concepts shape understanding of the world is at issue.

Challenging problematic metaphors

One way that the understanding of the objects of inquiry has affected knowledge is through the projection of gender. This sometimes occurs when metaphor is taken too literally and brings about a strong association of the object of inquiry with a particular gender. For example, feminist philosophers and historians of science have noted the use of gendered metaphors in research on reproduction. Emily Martin (1991), Bonnie Spanier (1995), and Evelyn Fox Keller (1996) have all described how gendered assumptions of stereotypically masculine traits for sperm and stereotypical feminine traits for the ovum misdirected research on reproduction through metaphorical descriptions that favored an active role for the sperm and a passive one for the ovum. In fact, the ovum plays an active role in reproduction and, even after this was observed by researchers, descriptions of ovum and sperm continued to reflect these stereotypes (Martin 1991).

Sarah Richardson discusses another example describing how a gendered assumption misdirected chromosomal research (2013). Thinking of X as the female chromosome and Y as the male chromosome was one of the factors in the formulation of the failed supermale hypothesis – the idea that XYY males have more "maleness" in virtue of having two Y chromosomes. "Having more maleness" was thought to include, among other things, being more aggressive. The reasoning went something like this: since Y distinguishes maleness, males with double Y should have more maleness. Since males are aggressive, double Y males will be even more aggressive. The conclusion turns out to be false, but the underlying assumption led not only to a false conclusion but also to a failure to examine evidence that did not support the hypothesis. While the greater number of XYY males who were in psychiatric institutions or prisons appeared to support the supermale hypothesis, the fact that these institutional populations also contain a disproportionate number of XXY males was over-looked. This disproportional presence of XXY males appears to be incon-sistent with the supermale hypothesis. The ultimate demise of the hypothesis was a result of other factors as well, however, the extension of gendered ste-reotypes to chromosomes is one example of how assumptions about gender can shape understanding of the object of inquiry and lead to the failure to consider relevant evidence. Richardson calls the projective gendering of the subject matter "gender valence" – a problem that is somewhat more subtle than blatant "gender bias".

Rebecca Jordan-Young and Katrina Karkazis (2019) document similar gender projection in the case of the hormones, testosterone and estrogen. Testosterone is typically identified as the male hormone and estrogen as the female hormone. This is in spite of the fact that both males and females produce testosterone and estrogen, something that was a surprise to early researchers. Jordan-Young and Karkazis point out that although these hor-mones are strongly associated with reproductive roles, they play a variety of other roles in human biology and the simplistic associations often made are problematic. For example, it appears that testosterone plays some role in ovulation and low testosterone in women can be a factor in reduced fertility. Furthermore, the strong associations often made between testosterone and male violence and testosterone and athletic prowess are not as straightfor-ward as they are often thought to be. High testosterone women (and men) are not always the best athletes, for example. Again, while the negative effects of the standard associations are felt most profoundly outside of sci-ence – in broader culture for example – researchers have not been immune from making assumptions that have adversely affected research based on these sorts of stereotypes.

Improving existing concepts

While metaphor is one way that conceptualization of objects of inquiry can affect research, it is not the only way. Concepts are analytical tools that serve to focus research on what matters for the inquiry at hand. A concept that works in one context may not always be transportable to different research contexts. One of the most crucial shifts relevant for feminism has been the conceptualization of "woman". Making a distinction between sex and gender contributed to the ability to interrogate social and political inequalities between men and women. The presumption of biological sex as "fixed", or permanent and determining of those inequalities, created barriers to doing so, but the contingency of culture and the understanding of gender as socially constructed allowed for questions to be raised about the role of women in society. While making this distinction served an important role historically, interdisciplinary research in the 1980s and 1990s, including analyses in feminist philosophy of science, supports that the biological category of sex is also socially constructed (Clune-Taylor 2021). Arguably, the very idea of nature and culture as dichotomous is a way of conceptualizing the world that may be problematic as well, and so, while the sex/gender distinction was useful in one context, it is no longer obvious that it is useful in all contexts. This is not an entirely pragmatic argument against the continued use of the distinction however, but rather a conclusion that is arrived at through concerns about the empirical, theoretical, and pragmatic evidence and the effects of making the distinction. That is to say, the body of knowledge that we have surrounding sex no longer supports the concept originally understood in the distinction.

The concept of democracy as used by political scientists provides a final example. Although the concept is complex and contested, there has been enough agreement about what democracy is to produce a robust subfield of democracy studies in political science. However, political scientist Pamela Paxton has offered a feminist critique of the standard formulations and measures of democracy. Paxton argues that most understandings of democracy treat universal suffrage as a key component of democracy but understand it, either implicitly or explicitly, as universal male suffrage. Paxton contends that the failure to distinguish full suffrage from male suffrage has both theoretical and empirical effects. If the understanding of "suffrage" is explicitly revised to include women's suffrage, then many of the key empirical and theoretical claims of democracy studies – including a standard categorization of the emergence of democracy in three distinct periods of history referred to as three waves – do not hold up (Paxton 2000). Paxton has been involved in a new project for conceptualizing democracy – the Varieties of Democracy Project (V-Dem). This project operationalizes the components of democracy in multiple ways, including a number of approaches that are sensitive to how gender differences affect regime classifications.

Recognizing and questioning the gendered nature of decisions about what counts as evidence is another way that what is known can be affected. Schiebinger discusses how archaeologists Margaret Conkey and Sarah Williams argued that the narrow understanding of tools, primarily as hunting tools, privileged accounts of culture revolving around the domination of nature rather than on social settings or family life, the former identified with the masculine and the latter with the feminine. Joan Gero challenged the assumption that tools were made by men, pointing to evidence that stone tools found in Peru and dating from 200 to 600 C.E. Peru were made by women (Schiebinger 1999:141).

Creating new concepts

The third way in which the investigation and understanding of the world can be affected by how the object of inquiry is conceptualized is when the phenomena of interest seem not to be covered by the standard terminology of the discipline, and so have no clear concepts associated with them. Phenomena without names are not conceptualized, cannot be theorized, and so cannot be understood. The coining of the term "sexual harassment" provides an example. The term was first used in the late 1970s to describe a variety of behaviors that women in consciousness-raising groups shared having experienced, particularly in the workplace. Although widespread, these patterns of behavior had not been understood as described under one concept. It was only when the similarity of experience emerged in these discussions that the need for a name – a concept under which to group the phenomena – became apparent (Farley 1978).

In a similar way, "intersectionality" was coined by Kimberlé Crenshaw (1989) in order to describe the form of discrimination experienced by Black women autoworkers. She argued that, in this case, discrimination was neither due simply to their gender nor to their race but rather resulted from their social location relative to both matrices of oppression. Although anti-discrimination statutes covered discrimination against women and Blacks, these women had lost their jobs because they were both Black and women – not because they were women and not because they were Black and so fell through the cracks of such anti-discrimination legislation.[5]

Within academic disciplines, topics that are not clearly defined and yet of interest might be thought of as inappropriate topics for research. Sociologist Marjorie DeVault describes the difficulty she had categorizing her research topic. She notes that it "is activity without a name, activity traditionally assigned to women; often carried out in family groups: activity that I know from experience but can not easily label" (DeVault 1991: 4). She settles on the phrase "feeding the family" to group together a related variety of tasks: planning meals, shopping, cooking, serving, cleaning up, and others. She notes, "This particular insufficiency of language is an example of a more general problem, a more pervasive lack of fit between women's experiences and the forms of thought available for understanding experiences" (DeVault 1991: 5). Redressing such inabilities to capture the realities of lived experience is another way that feminist thought may alter what is known.

In one more example of the insufficiency of traditional approaches, Sari van Anders has developed a framework for the study of sexuality that she calls Sexual Configuration Theory, (SCT). Van Anders notes that standard measures used for investigating sexuality (such as the Kinsey Scale and the Klein Sexual Orientation Grid) have as a primary focus sexual orientation and are limiting for this reason. Specifically, "A major limitation of existing theories about sexual orientation is that they do not always map onto people's actual experiences, limiting these theories' validity" (van Anders 2015: 1179). Part of the reason for this is that they limit investigations of sexuality through assumptions that are made about the alignment of gender and sex and what is included in each of these concepts. Van Anders proposes SCT as a framework that considers a variety of factors relevant to sexuality that other measures do not consider. "There are a number of other axes along which sexuality could revolve, including age, partner number, type of sexual activity, consent, solitary sexuality, and intensity among others" (van Anders 2015: 1178). She argues that current understandings of sexual orientation are problematic in a variety of ways. They fail to acknowledge sexual fluidity – shifts in sexual attraction and non-exclusivity, for example. They also treat sexuality as dyadic, and so treat partnered and solitary sexuality as overlapping, with the latter being the substitute for the former, an unexamined assumption. In addition, they treat partnered sexuality in terms of sameness and difference (homosexuality vs heterosexuality) without taking into consideration that what counts as "the same" is variable. A further difficulty of this binary orientation is that it does not accommodate the sexualities of many who are genderqueer or trans.

SCT reconceptualizes sexual orientation by providing an alternative framework (sexual configuration). It makes use of diagrams that visually represent the wide variety of parameters van Anders identifies in order to measure sexuality (individuals locate themselves using the diagrams). Van Anders offers "sexual configuration" as an alternative to "sexual orientation", choosing "configuration" because "the term denotes dynamism rather than fixedness" (van Anders 2015: 1185). It is a feminist approach in that its source is the desire to capture the lived experiences of those who are studied (see the discussion of standpoint theory in Chapters 2 and 4), but also in its recognition that marginalized sexualities could not be studied with the frameworks that were currently being employed.[6]

Each of the examples in this section illustrate ways in which changes in understandings of the objects of inquiry improved knowledge production either through opening a new area of research or through transforming an already existing research area, so that it was better suited to produce knowledge that supported both the epistemic and non-epistemic (feminist) goals of research.

Whether or not it is legitimate to conceive of the objects of inquiry in the alternative ways that any account proposes depends on a variety of factors. As we have seen, the motivation for proposing alternatives is often the empirical inadequacy of those currently available. Paxton notes that nation-states identified as democratic do not actually meet the standards they claim to. DeVault can find no description of the realities that she experiences and wants to study. Van Anders observes that the lived experiences of sexuality cannot be captured by current measures. There may also be theoretical motivations for conceptual change. Van Anders notes the theoretical sources of SCT in bioscience (van Anders 2015: 1187). But a third mode of assessment for the viability of conceptual change is the usefulness of the concept for achieving the goals of knowledge production. The adequacy of the concept and the consequent ontological objectivity of an account that makes use of it depends on its empirical adequacy, its theoretical coherence, and its pragmatic success. We close the chapter with a brief discussion of what we mean by suggesting these three constraints on objectivity.

Conclusions

In this chapter we have distinguished ontological and epistemological aspects of objectivity. The former has to do with *what* it is that we know when we know objectively whereas the latter is concerned with *how* we know objectively. Most feminist philosophers of science have been interested in maintaining some understanding of objectivity, in part, because a claim that feminist science is better science would seem to require it. We have described two ways in which the increasing participation of women in science has changed the objects of inquiry. The scope of inquiry was expanded as women sought knowledge of topics that had been ignored, under-studied, or overlooked. In addition, feminists have reconceptualized objects of inquiry so that they no longer reflect sexist assumptions, better reflect lived experience, and are better able to address the research questions that were of interest to them. In this way, they have altered the ontology of inquiry. We have offered examples of each of these ways in which feminist work has affected the sciences.

What we have not done in this chapter is address the question of how the changes in what we know, brought about through feminist work, have improved the objectivity of science. Most of the efforts of feminist philosophers of science and epistemologists directed to discussion of objectivity have focused on the epistemological aspects of objectivity – *how* we know more objectively. We turn to a discussion of those in the next chapter. Here we offer some suggestions for how to think about ontological objectivity given what we have said about feminism and the objects of inquiry.

Remember that we described ontological objectivity as capturing the idea that we have gotten something right about the world. When accounts of the objects of inquiry get something right, they pick out features of reality and those features are significant in the sense that they address the aims of research. Yet, even though there might be multiple aims that focus on different features of reality, there are still empirical constraints on the objects of inquiry. They are empirically constrained in that there are experiences that can support or contradict particular accounts. Heather Douglas (2004) also points to convergence – different lines of inquiry leading to the same result – as a potential indication that empirical constraints have been met. When accounts of the objects of inquiry follow from different paths of investigation and come to the same conclusion – or converge on the same result – this may be evidence that an account is ontologically objective. Additionally, ontological objectivity might require that any understanding of objects of inquiry should converge with other accepted theoretical knowledge or, if not, it must be possible to explain why not. We refer to this as accounts being theoretically constrained. Finally, we have suggested that what is taken to be significant or relevant shapes how objects of inquiry are conceptualized or represented and depends on the aims of the research. Some of these are epistemic aims of course, but feminist research also includes the political, social, and cultural aims of feminism. The objects of research are thus also pragmatically constrained. If they do not allow us to successfully address the aims of the research – both epistemic and other – they are inadequate. To summarize, the objectivity of knowledge requires that the objects of inquiry are empirically, theoretically, and pragmatically constrained. When we say that an account is ontologically objective, it is an account that gets something right about the world and the determination of whether it does so involves assessing the extent to which it is empirically, theoretically, and pragmatically constrained.

This is far from a complete feminist analysis of the ontological aspects of objectivity, but it offers some suggestions for how to develop one – one that shows that feminist epistemology, in particular the claim that knowledge is situated or depends on interests, can be compatible with maintaining that what we have knowledge of is "objective" or not merely relative to, or a product of, human beliefs. The epistemological aspects of objectivity are much more fully developed in feminist work. We explore these in Chapter 4.

Discussion questions

1. What is the "value-free ideal"? Why does this ideal seem plausible, especially for science?
2. What are the three senses of objectivity that are identified in this chapter? Why might some conception of objectivity be important, especially in science?
3. The text makes a distinction between the objects of inquiry and the objects in the world. What is this distinction and to what use is it put?
4. What are the ways in which feminists changed what we know (either through changing the scope of inquiry or changing how we think about those objects)?

Additional suggested readings

Ashton, N. (2019). Relativising epistemic advantage. In: M. Kusch (ed.) *The Routledge handbook of philosophy of relativism*. New York: Routledge, pp. 329–338. An argument that standpoint theory is committed to a non-pernicious form of relativism.

Crasnow, S. (2008). Feminist standpoint theory. In: N. Cartwright and E. Montuschi (eds.) *Philosophy of social science*. Oxford: Oxford University Press, pp. 145–161. Primarily about standpoint theory but it closes with an account of interest-based objectivity revolving around the conceptualization of objects of inquiry.

Crasnow, S. (2021). Coherence objectivity and measurement: The example of democracy. *Synthese*, 199, pp. 1207–1229. An account of the ontological aspects objectivity through empirical, theoretical, and pragmatic constraints on conceptualization.

Intemann, K. (2021). Feminist perspectives on values in science. In: S. Crasnow and K. Intemann (eds.) *The Routledge handbook of feminist philosophy of science*. New York: Routledge, pp. 201–215. Situates feminist considerations about values in science in the general discussion of the issues in philosophy of science.

Rolin, K. (2021). Analytic feminist approaches. In: K. Q. Hall and Ásta (eds.) *The Oxford handbook of feminist philosophy*. Oxford: Oxford University Press, pp. 226–236.

Scheman, N. (2011). *Shifting ground: Knowledge and reality, transgression and trustworthiness*. Oxford: Oxford University Press.

Notes

1 The three senses of objectivity here are consistent with the three senses that Douglas (2004) identifies, though we use a simplified version of that account here. Readers who are interested in finer-grained distinctions that might be made under these three general categories should see Douglas 2004.

2 Those interested in the bias paradox and how feminists have responded, see, for example, Campbell 1998; Heikes 2004; Rolin 2006; Intemann and de Melo-Martín 2016; Intemann 2017; Engqvist 2022.

3 We write here as if we were describing all women entering scientific fields during this period and, of course, this is false. It is important to keep in mind that women as a group do not all have the same interests or goals and we do not intend to claim that they do. However, to write continually that we are speaking of some women is awkward and our goal is to describe general trends during this period and so, with apologies, we will continue to use the general formulation except where it is crucial that differences are considered.

4 Recent discoveries at a burial site in Peru further support that women played a role as hunters and were more involved in subsistence activities than earlier paleoanthropologists had believed (Anderson et al. 2023).

5 Specifically, these women lost their jobs because they were last hired and so lacked seniority. But they were last hired because they were Black women. The law said that they could not be discriminated against because of race or gender – that is, it had to be shown that they had been hired last either because they were women or because they were Black. But White women had been hired earlier (showing that women were not being discriminated against) and Black men had been hired earlier (showing that Blacks were not being discriminated against). But their late hiring was due to the intersection of their race and gender (Black women were the last hired). Crenshaw argued that the law did not account for such intersectional discrimination and yet it clearly existed.

6 It is possible to describe what van Anders is doing is revising the concept "sexual orientation", but in fact she retains sexual orientation as one parameter for sexual configuration. This supports that sexual configuration is intended as a new concept (van Anders 2015:1182–1183).

References

Anderson A, S. Chilczuk, K. Nelson, R. Ruther, C. Wall-Scheffler. (2023). The myth of man the hunter: Women's contribution to the hunt across ethnographic contexts. *PLoS ONE*, 18 (6), e0287101. https://doi.org/10.1371/journal.pone.0287101.

Barker, D., and E. Kuiper. (2021) Feminist economics. In: S. Crasnow and K. Intemann (eds.) *The Routledge handbook of feminist philosophy of science.* New York, NY: Routledge Press, pp. 355–367.

Campbell, R. (1998). *Illusions of paradox: A feminist epistemology naturalized.* Lanham, MD: Rowman & Littlefield Publishers.

Clune-Taylor, C. (2021). Is sex socially constructed? In: S. Crasnow and K. Intemann (eds.) *The Routledge handbook of feminist philosophy of science.* New York, NY: Routledge Press, pp. 187–200.

Crenshaw, K. (1989). Demarginalizing the intersection of race and sex: A Black feminist critique of anti-discrimination doctrine, feminist theory and antiracist politics. *University of Chicago Legal Forum*, pp. 139–167.

Daston, L., and P. Galison. (1992). The image of objectivity. *Representations*, 40, pp. 81–128.

Daston, L., and P. Galison. (2007). *Objectivity*. New York: Zone Books (MIT Press).

DeVault, M. (1991). *Feeding the family: The social organization of caring as gendered work*. Chicago: University of Chicago Press.

Douglas, H. E. (2004). The irreducible complexity of objectivity. *Synthese*, 138, pp. 453–473.

Engqvist, T. (2022). The bias paradox: Are standpoint epistemologies self-contradictory? *Episteme*, 19 (2), pp. 231–246.

Farley, L. (1978) *Sexual shakedown: The sexual harassment of women on the job*. New York: McGraw-Hill.

Hacking, I. (1999). *The social construction of what?* Cambridge, MA: Harvard University Press.

Heikes, D. K. (2004). The bias paradox: Why it's not just for feminists anymore. *Synthese*, 138, pp. 315–335.

Intemann, K. (2017). Feminism, values, and the bias paradox: Why value management is not sufficient. In: K. C. Elliott and D. Steel, *Current controversies in values and science*. New York: Routledge, pp. 130–144.

Intemann, K., and I. de Melo-Martín, (2016). Feminist values, commercial values, and the bias paradox in biomedical research. In: M. C. Amoretti and N. Vassallo (eds.), *Meta-philosophical reflection on feminist philosophies of science*, New York: Springer, pp. 75–89.

Janack, M. (2002). Dilemmas of objectivity. *Social epistemology*, 16 (3), pp. 267–281.

John, S. (2021). *Objectivity in science*. Cambridge, UK: Cambridge University Press.

Jordan-Young, R., and K. Karkazis. (2019). *Testosterone: An unauthorized biography*. Cambridge, MA: Harvard University Press.

Keller, E. F. (1996). *Refiguring life: Metaphors of Twentieth-Century biology*. New York, NY: Columbia University Press.

Kitcher, P. S. (2001) *Science, truth, and democracy*. New York, NY: Oxford University Press.

Lloyd, E. (1995). Objectivity and the double standard for feminist epistemologies. *Synthese*, 104, pp. 351–381.

Martin, E. (1991). The egg and the sperm: How science has constructed a romance based on stereotypical male-female roles. *Signs*, 16 (3), pp. 485–501.

Paxton, P. (2000). Women's suffrage and the measurement of democracy: Problems of operationalization. *Studies in comparative international development*, 43, pp. 92–111.

Richardson, S. (2013). *Sex itself: The search for male and female in the human genome*. Chicago, IL: Chicago University Press.

Rolin, K. (2006). The bias paradox in feminist standpoint epistemology. *Episteme*, 3 (1–2), pp. 125–136.

Schiebinger, L. (1999). *Has feminism changed science?* Cambridge, MA: Harvard University Press.

Spanier, B. (1995). *Im/partial science: Gender ideology in molecular biology*. Bloomington, Indianapolis: Indiana University Press.

Tuana, N. (2006). The speculum of ignorance. *Hypatia: A Journal of Feminist Philosophy*, 21 (3), pp. 1–19.

van Anders, S. M. (2015). Beyond sexual orientation: Integrating gender/sex and diverse sexualities via sexual configurations theory. *Archives of Sexual Behavior*, 44, pp. 1177–1213.

Washington Post. (2021, April 19). Jurors start deliberation. https://www.washingtonpost.com/nation/2021/04/19/derek-chauvin-trial/ (accessed July 14, 2022).

4

HOW DO WE KNOW?

Overview

Both the view that knowledge is achieved through epistemically reliable methods that could be employed by anyone (methodological objectivity) and the idea that epistemic agents should pursue knowledge in detached ways or free of any interests or values (agent objectivity) are legacies of the traditional understanding of how we know that embraces the value-free ideal (VFI). These are epistemological aspects of the concept of objectivity that have governed accounts of *how* we know. This chapter focuses on how gender norms and relations may impact *how* we know and how feminists have approached the epistemological aspects of objectivity while rejecting the VFI. We identify nine phases of scientific inquiry and consider how the choices that are made during each of those phases affects the knowledge produced. Using examples from a variety of disciplines, we show that values and social norms play a role in which methodologies are employed, what we observe, how data is described and interpreted, as well as judgments about the *kind* of evidence needed and *how much* evidence is needed. We consider several different feminist approaches to epistemological questions about the objectivity of knowledge, including Helen Longino's contextual empiricism, Sandra Harding's notion of "strong objectivity", and Alison Wylie's analysis of objectivity. We also consider some pragmatists approaches to adjudicating values in science.

Introduction

Examining whether there are sex differences in the brain or cognitive abilities has been the focus of research for centuries (Gould 1996). Donna Maney

DOI: 10.4324/9781032693781-4

(2016) discusses how James Crichton-Browne's research on sex differences in the brain were communicated in the *New York Times* in 1912. Crichton-Browne, a prominent neuropsychologist and collaborator of Darwin, purported to find that differences in blood flow in male and female brains gave women greater abilities for "sensuous perception, rapidity of thought, and emotional sensibility" whereas men were more likely to have "greater originality on higher levels of intellectual work", "calmer judgment", and "stronger will" (*New York Times* 1912). While the "blood flow" theory gave way to more nuanced and complex explanations (such as genetic differences, hormonal differences, or other structural or neurological differences), the idea that there are cognitive sex differences that are biologically determined has persisted. How is knowledge about sex differences produced? What sort of empirical evidence is collected and how does it support conclusions about sex differences? Is the knowledge produced objectively or in ways that may be limiting, biased, or problematic? While the previous chapter focused on how gendered values, norms, and the social aspects of knowers might influence *what* we know, this chapter addresses whether such considerations can influence *how* we know. We examine how gender can influence experimental design, methodology, interpretation of data and inferences to particular conclusions. As noted in Chapter 3, this might appear to be in tension with traditional accounts of methodological objectivity (the idea that knowledge is acquired through methods that are epistemically reliable and could be employed by anyone) and agent objectivity (the idea that those producing knowledge should be free of any idiosyncratic values or biases). Here we consider how feminist critiques of methodologies contributed to the development of alternative accounts of objectivity and recommendations for minimizing problematic gender biases. We will also show how these alternative conceptions allowed feminists to uphold objectivity while acknowledging a variety of ways in which values and interests can play inevitable and sometimes beneficial roles in producing knowledge, particularly in science.

The complexity of scientific methodologies and reasoning

While we often learn that there is a "scientific method" that involves generating a hypothesis, testing that hypothesis, and arriving at a conclusion; the reality is more complicated and different areas of science employ a wide range of methodologies, experiments, kinds of reasoning, and – as we saw in the last chapter – conceptual frameworks. Nonetheless, it will be useful to distinguish different phases of research that contribute to how researchers arrive at scientific knowledge (broadly speaking). Each of these phases involves making choices about how we investigate the objects of inquiry and produce knowledge. While scientific inquiry might be carved up in a number of ways, we offer the following characterization of the phases of inquiry (see also Anderson 2004 and Brown 2020 for similar characterizations):[1]

1.   Choosing a research area/subject
2.   Framing of the questions to be addressed about that subject (hypothesis generation)
3.   Deciding which observations are relevant
4.   Collecting data based on those observations
5.   Analyzing or interpreting that data
6.   Deciding when to stop data collection
7.   Drawing conclusions based on the evidence
8.   Considering alternative hypotheses or explanations for the data
9.   Testing those conclusions for robustness (including evaluation of alternative hypotheses).

In practice, these phases of research can be messy to distinguish because, as we will see, they can influence each other. This is because the process of knowledge production is iterative – decisions made at one phase of the inquiry may be revised in light of results at another phase. Distinguishing between them conceptually, however, is useful for understanding how decisions that researchers (and all knowers) continually make throughout knowledge production might be influenced by gendered values, norms, assumptions, or other social factors. As we will see, such choices in practice may undermine traditional conceptions of methodological and agent objectivity.

In Chapter 3, we focused on the ontological notion of objectivity – objectivity about what is "really real" – and saw how values, norms, assumptions, and the social characteristics of researchers or research contexts can limit or shape the objects of inquiry. We showed that this can influence *what* we know. The examples and discussion in Chapter 3, however, also demonstrate how such values and norms can influence *how* we begin to know during phases 1–3: choosing a research subject, framing research questions, and deciding which features of a phenomenon are relevant to observe. These first three phases involve judgments about what objects to study, how to conceptualize them, and which of their features are relevant or important to an area of inquiry. In research on sex differences in the brain, interests in understanding (and perhaps justifying) perceived differences and inequalities between men and women drove much of this research (phase 1) and research questions (phase 2) were framed around the assumption that any behavioral differences could be explained by essential biological differences. Such assumptions also influenced what was relevant to observe (phase 3). It was assumed that brains come in two kinds (male and female) and that observed biological differences in the brain would explain differences in abilities.

This chapter will focus primarily on how feminist work has demonstrated that gendered biases can enter decisions related to collecting and interpreting evidence (phases 4–9) and the implications that this might have for methodo-logical and agent objectivity. Of course, as noted, these phases can interact. For example, differences in how the object of inquiry is conceived will affect what is considered relevant (phase 3) and therefore what counts as evidence in later phases. In addition, what is considered relevant may also be affected by the choices that researchers make about their methodology, as will become clear in what follows.[2] Nonetheless we will focus on the later phases, parti-cularly as these may be more controversial.

Gender and the phases of research

The practices of scientific inquiry, the norms determining appropriate meth-ods, and the procedures that are used by the scientific communities doing research affect the choices that are made about data collection, testing proce-dures, and conclusions that are warranted based on the evidence. The question of whether specific claims are open to further investigation or scrutiny and how that investigation should be conducted are questions that revolve around such practices and procedures. Similarly, the question of how much and what kinds of evidence are sufficient to justify acceptance or belief in a claim is also procedural. When we say "practices" and "procedures" here, we mean both the techniques that are accepted as appropriate for producing knowledge within scientific communities and the practices that are both explicitly and implicitly understood by such communities to be the "right sort" for produ-cing, assessing, and inferring from the evidence produced through those techniques.

To illustrate, consider the widely studied area of potential cognitive sex difference in spatial abilities. One experiment that was eventually applied in such research was originally developed in 1948 by Witkin and Asch to dis-tinguish those who are "field-dependent" perceivers (who rely on the sur-rounding context and cues to determine how objects are oriented) and "field-independent" observers (who can determine how things are oriented inde-pendently of their surroundings). In the original version of this experiment, a subject is placed in a completely dark room and sits in a chair that the experimenter can tilt to different angles. Once in the chair, the subject is presented with two illuminated objects: a rod and frame (each of which are also tilted to different angles). The subject is then supposed to instruct the experimenter on how to move the rod so that it lines up to perpendicularly bisect the tilted frame. Subjects who can successfully align the rod to bisect the frame, regardless of how the frame and their own chair are tilted are thought to be more "field-independent" and have stronger visual-spatial skills. In the 1950s to 1970s, researchers began employing versions of this experiment to test whether men had better visual-spatial skills than women (see Macoby and Jacklin 1974 and Fausto-Sterling 1985). Sure enough, they found that men tended to perform better on the rod and frame test (they were more likely able to align the rod to bisect the frame regardless of how they or the frame were tilted). Researchers concluded that men had superior visual-spatial abilities. Indeed, versions of this test are still used today (often in conjunction with other tests – such as the hidden figures test) to conclude that there are sex differences in spatial abilities (e.g., Yuan et al. 2019; Abdul Razzak and Bagust 2022).

How does this relate to the phases of research identified above? In order to test a hypothesis (that men have better visual abilities), scientists designed an experiment to generate observations. They assumed that the rod-frame test was a relevant measurement of visual-spatial abilities (phase 3). They collected data that men (or boys) were able to correctly align the rod to a greater degree than women (or girls) (phase 4). They interpreted that data as supporting the claim that men had better cognitive spatial abilities than women (phase 5). Researchers thought there was sufficient data – as similar results were found by multiple studies and also cohered with evolutionary theories about why men and women might have different cognitive and physical differences (phase 6). They concluded such differences were likely best explained by biological differences between men and women (phase 7).[3]

On the surface, one might think that this sort of research upholds standards for what we might consider "right methods" – at least in phases 3–7 of research as identified earlier. Of course, the conclusion might make some uncomfortable, but that by itself does not mean that the research is unsound or biased in some way. Feminists, however, have criticized research using the rod and frame test, as well as a variety of other studies that purport to show that there are biologically determined cognitive sex differences that track certain abilities.

Feminist and biologist Anne Fausto-Sterling presented one of the first detailed feminist critiques of the research that had been done on cognitive sex differences in her 1985 book, *Myths of Gender*. As Fausto-Sterling points out, there were several flawed assumptions and methodological choices, particularly in studies that used the rod and frame test. It is not entirely clear what the rod and frame test measures. The test seems to rely on the ability of human subjects, in a dark room, to be comfortable and assertive enough to instruct a (typically male) experimenter to make tiny adjustments to the lighted rod numerous times until it is "perfect". It probably comes as no surprise to us today that during the 1950s–1970s, women may not have been comfortable being assertive or demanding of male researchers to continually get them to make the necessary adjustments (particularly while in a dark room and being tilted in different positions). But this is not necessarily because there was something different about women's brains. It is quite likely a reflection of gender norms and roles at this time, where women were encouraged to not be too demanding or do anything that might "make more work" for an experimenter. Not to mention that some women were likely to be uncomfortable in a dark room alone with a male experimenter and might not perform to the best of their abilities. In this case, decisions about what observations were relevant (including the decision about how those observations would be made) perhaps incorrectly assumed that there were no other non-experimental factors influencing how men and women would perform on this test. A lack of attention to other gendered social factors potentially shaped and limited what data was collected.

In addition, researchers then (and many who followed) also likely shared a widely held assumption that men and women did in fact have differences in cognitive abilities and skills – and this widespread assumption made it quite easy to interpret the data as supporting that, as opposed to some other explanation. Indeed, this widespread assumption likely also influenced decisions about how much data was collected (phase 6). When the result is not as surprising, one might think that less evidence is needed to justify a conclusion. In deciding when we can stop collecting data – when the evidence is *sufficient* to justify a particular conclusion – scientists implicitly make a judgment about how much evidence is enough, given the risks of error (Douglas 2009). Scientists want to avoid accepting too quickly a hypothesis that is false or rejecting too quickly a hypothesis that might in fact be true. Yet, how much evidence is enough depends not only on the probability that a hypothesis is true or false given the evidence, but also what the consequences of error would be. This is known as inductive risk.

What the risks of error are, however, depends on value judgments (Rudner 1953; Douglas 2009). For example, if we falsely believe that there are biologically determined sex difference in spatial skills, this would erroneously reinforce gender stereotypes that might impact certain kinds of policies and interventions (for example, encouraging and supporting young girls in areas or professions where those skills might be relevant such as math, engineering, and architecture). This runs the risk of discouraging a large section of the labor market from interests in such professions and may also hinder women's economic prospects insofar as those are highly paid professions. Moreover, the probability of reinforcing such stereotypes is increased, given that they are widespread and we tend to give more weight to evidence that coheres with our existing beliefs (Kitcher 2001).

A further risk of this research is that it reinforces not only gender stereotypes, but a two-sex model that may not even be empirically adequate (Fausto-Sterling 1993). Similarly, rejecting the biological determination view of spatial abilities too quickly, if it is in fact true, might also carry certain risks. That is, we might erroneously deprive ourselves of better understanding differences and how best to address them in just ways. We may also adopt misguided policies that have little chance of success. Which of these errors is more acceptable is a value judgment (about which would be worse). When the risks of error are significant, it is often more responsible to collect more data, especially if action is not urgent, or at least avoid making sweeping conclusions about what the data provides evidence *of*.

In this case, however, researchers often did make rather grand conclusions about sex differences and the significance of their findings (phase 7) (Fausto-Sterling 1985). They concluded that differences in performance on the rod and frame test could *only* be explained by biological sex differences that determined visual-spatial abilities. This seems unwarranted because, as Fausto-Sterling points out (and other studies have since shown), the actual differences are quite small, even though statistically significant, such that it is not clear what the difference would amount to in relation to one's ability to perform certain tasks or activities in everyday life. Thus, values shaped the ways in which researchers drew conclusions and particularly how they characterized the significance of the conclusions by focusing on the fact that there was a statistically significant difference (as opposed to focusing on the small degree of that difference).

Researchers in this case also appear to have overlooked several other alternative hypotheses (phase 8). As mentioned, such differences may be the result of complicated environmental and social factors that led to the gendered differences in abilities. Or it might be explained by experimental error – that the rod and frame experiment better tested women's ability to be assertive and picky during the experiment than visual-spatial skill. It was not necessarily that the researchers in this case had bad intentions or were determined to show that women were inferior, but rather various alternative hypotheses that could also explain the data simply never occurred to them. Values and experiences (shaped by social position) can also influence the kinds of hypotheses that we think of, or the kinds of explanations that might occur to us. Yet, insofar as drawing conclusions requires making inferences to the best explanation, then such values may influence how strongly our conclusions are in fact supported as the *best* explanations (Okruhlik 1994; Elliott and McKaughn 2009).

In this case, the alternative hypotheses that went unconsidered point to various confounding variables, or factors other than the one being studied (biological sex differences) that may be associated with the behaviors of interest (visual/spatial performance). Such confounding factors can distort or mask the effects of other variables on the behaviors in question. When researchers identify these variables, they can try to design experiments to better control for them. This is important not only for ensuring good initial experimental design (phase 3 and 4) but also to test and evaluate initial findings for robustness (phase 9). Indeed, later studies that did try to establish further controls and correct for some limitations with the rod and frame test found significantly less difference in visual-spatial performance between men and women (see for example, Richardson 1994).

Of course, in hindsight, the tests that were done during the 1950s–1970s might be dismissed as cases of "bad" science, where scientists failed to notice experimental design problems or alternative hypotheses, interpreted the data according to their preconceived notions, and drew hasty conclusions that were not properly attentive to the practical significance of their findings. Yet, these were respected scientific studies in their time, published in peer-reviewed journal articles, and still influence a large field of research on cognitive sex differences. Moreover, it seems as though some of the ways that values entered here (such as how they influence inductive risk calculations, or the kinds of alternative hypotheses we consider) are inherent to all instances of knowledge production in a way that perhaps cannot be easily addressed (and is likely to occur in much of science). But if, contrary to the value-free view, values play a necessary role in some of the decisions about evidence, what implications does this have for scientific objectivity, or how we might minimize bias in science? This will be the focus of the next section.

Objectivity and method

If gendered experiences, assumptions, and values can enter or influence the phases of research described above, what implications does that have for how we can know objectively? Understanding objectivity as the "right method" (methodological objectivity) is problematic if we think of it as simply following scientific methods because, as the above example shows, values, assumptions, and social experiences can influence how scientists make decisions in collecting and interpreting data. Moreover, it is not clear that the subjective features or biases of individual knowers or communities can be stripped away or prevented from influencing such decisions (agent objectivity).

Of course, one might conclude at this point that science (or knowing more generally) simply is *not* objective. But that response seems unsatisfactory (and not one that many feminists have wanted to endorse). In some sense, we want to be able to say that science can do better, or that in some cases the science or knowledge that is produced is less distorted or more reliable, complete, or useful. Thus, feminists have re-framed the traditional question "how do individuals objectively know" to "what conditions must be met for knowing to be objective (in some sense)?" For example, might there be procedures, processes, and practices that could help counteract the negative influences of individual biases and limitations?

Feminist philosophers of science and epistemologists have taken a variety of different approaches to this question. Some have argued that better implementation of the standard empiricist methods of science can eliminate such biases. In other words, they have suggested that it is when science is not practiced properly that bias distorts knowledge. Sandra Harding described feminist epistemologists who take this approach as "spontaneous feminist empiricists" (Harding 1986). Those adopting this approach affirm that properly following (objective) procedures, processes, and practices will produce objective knowledge because they ensure that values never function as evidence. In this way, the sexist values that feminists identify as distorting knowledge can be eliminated or at least relegated to roles that are thought not to affect the conclusions reached. This position may be consistent with traditional assumptions of agent and methodological objectivity. Problems occur when researchers allow personal biases to influence their work, such that empirically reliable methods are not correctly applied.

But, as we have seen in the previous section, sometimes practices that are thought to constitute good science may incorporate values implicitly, so that they are not recognized and produce biased results. In addition, some feminist epistemologists (along with other philosophers of science) have argued that values are ineliminable since they play a role in the choices researchers make through the nine phases of research. If so, then rather than attempting to eliminate them we need to adopt procedures, processes, and practices that support the legitimate use of values while preventing their illegitimate use. To do so requires being able to make a distinction between legitimate and illegitimate roles for values. It could be, for example, that values are crucial for choosing research topics and framing of relevant questions about those topics (phases 1 and 2), but are possibly problematic during any of the other phases. Phases 3–7, in particular, involve choices that revolve around evidence. This distinction between phases may be difficult to maintain, however, insofar as the phases interact and can operate in an iterative (not linear) manner. We have already seen, for example, how decisions in phases 1 and 2 (what is studied and how research questions are framed) influence decisions in phase 3 (what evidence is relevant), which in turn influences what evidence there is.

One area in which feminists have differed is over the question of how values might be relevant to evidence, including whether values might be evidence themselves (see the section "Feminist holist empiricism: The pragmatist solution" in this chapter). If so, the VFI that undergirds traditional empiricism and spontaneous feminist empiricism is no longer attainable and any notion of objectivity that depends on that ideal must either be abandoned or revised. Most feminist philosophers of science and epistemologists have wanted to retain objectivity as a virtue of science and of knowledge in all its forms, and consequently they have offered alternative accounts of what makes for good (objective) knowledge production. We explore some of these accounts in the remainder of the chapter.

The contextual empiricist solution: Procedural objectivity

Helen Longino proposes an account of evidence that acknowledges the way that conclusions are dependent on the background assumptions that shape the research context. She begins with an account of evidence in which phenomena only become evidence relative to some set of background assumptions (1990, 2002) which indicate that these phenomena are relevant to the question under investigation. She calls this account "contextual empiricism" since it is empiricism – only empirical phenomena count as evidence – and it is contextual, in part, because empirical phenomena become evidence only when their relationship to theory is established through the appeal to various background assumptions and knowledge that are part of the context of research. Different contexts might include different background assumptions (including value assumptions). Empirical phenomena that would be evidence under one set of background assumptions might not be evidence in a context in which those assumptions are not held.

Consider an example from feminist economics. Julie Brines (1994) challenged the standard economic explanation of the unequal distribution of housework in heterosexual households. Roughly, that explanation is that because women typically earn lower wages, they have less bargaining power, and so are unable to negotiate a "better deal" on housework, resulting in most domestic tasks falling to them. This explanation fails to account for the fact that it is also the case that when women have higher earnings than men, the balance does not shift. The data she cites from the period (1990s) indicates that men did even *less* housework when their female partners earned more than they did. This data is not evidence for or against the hypothesis that higher wages for men result in less bargaining power for women, although it certainly raises questions. Shifting the context so that it includes a gender analysis of heterosexual marriage as an arena for gender performance (not merely an arena for economic bargaining) reveals the relevance of the data as an anomaly to be explained and suggests an emendation to the original hypothesis. The background assumption that only economic factors, narrowly understood, were relevant limited what empirical phenomena could be considered relevant.

Brines argues that the combination of the gendered identification of housework with women and the greater earning power as gender identified with men produces a need for men who were out-earned to perform gender within their marriages by not engaging in housework. In other words, she proposes that what explains men doing even less housework when they are out-earned is that they are compensating for their lack of performance as earners. Understanding that the amount of housework done is connected to gender role performance, and not fully accounted for by an economic model that does not take gender into consideration, requires different assumptions. It requires, for example, recognizing that there are beliefs about the nature of marriage and the power dynamics within heterosexual marriage that are built into the standard account. These assumptions are questioned and revised in Brines's account and the feminist value that gender is relevant guided her analysis. Differences in background assumptions and values – part of the context of research – can change what is taken as evidence. That men do even less housework when out-earned is not part of what needs to be explained in the standard account, but for Brines it is evidence and can only be understood as evidence when the relevance of gender is acknowledged. As in this case, such assumptions often depend on social, political, and cultural values – contextual values. Consequently, contextual values play a role in scientific reasoning, making evidence relative to the context.

The Brines example is interesting as well since it illustrates how shared assumptions within a scientific community (that gender is not relevant to power dynamics within a marriage or alternately, that marriage is just another venue for economic analysis) may go unnoticed and unchallenged. The feminist values that Brines brings to her research allow her to notice and challenge those assumptions. Values affect her choice of research question (phase 1), which observations she takes to be relevant (phase 2), her interpretation of data (phase 5), and lead to her consideration of alternative hypotheses (phase 8). Conversely, the account that she is challenging has made different choices at each of the phases.

Elizabeth Anderson's analysis of Stewart et al.'s research on divorce (discussed in Chapter 2) provides another example. Reports of the various ways in which their lives had changed after divorce, including accounts of improved wellbeing, were not considered (or even sought) as evidence in previous research because of value assumptions about the negative character of divorce.

Longino's contextual empiricism offers an account of evidential relevance that illustrates how social, political, and cultural values are implicated in the procedures, processes, and practices of knowledge production. Thus far we have not discussed how she addresses the question of objectivity, however. Longino provides an alternative understanding of objectivity based on the social nature of knowledge. Given that the context of knowledge is social, the way to root out problematic assumptions is also social. Longino proposes what Janet Kourany has dubbed the "social value management ideal" (Kourany 2010). Rather than taking the social character of science to be problematic, Longino sees it as providing grounds for objectivity by creating the opportunity for "transformative criticism".

Transformative criticism is possible within scientific communities that conform to the following four norms: 1) the existence of publicly recognized forums for the criticism of evidence, methods, assumptions, and reasoning; 2) uptake of criticism, which requires that the community not merely tolerate dissent, but that it alters its beliefs and theories over time in response to the critical discourse; 3) publicly recognized standards for the evaluation of theories, hypotheses, and observational practices that are appealed to in order to make and respond to criticism; and finally, 4) communities should be characterized by tempered equality of intellectual authority, which requires taking seriously those with the appropriate expertise and not dismissing their criticisms on the basis of irrelevant features (like gender, race, or class) (Longino 1993, 2002). These norms allow for the management of individual biases or partialities which may be involved in background assumptions either explicitly or implicitly, because they provide means of making values explicit and open to critical examination.

Science is not value free but rather the values that are present in scientific inquiry can be (socially) managed when inquiry conforms to these four norms. Contextual values can be adjudicated in public forums when values conflict. They are open to critical examination to ensure that they do not create unwarranted bias and the public nature of the venues for criticism means all value perspectives can be represented in such forums. Uptake calls not only for criticism to be allowed but for criticism to be responded to. However, that criticism must itself meet certain standards – again publicly recognized. Finally, equality of intellectual authority allows for the participation of all value perspectives, but it is "tempered" to recognize the differential distribution of expertise within communities.

The requirement that knowledge communities have equality of intellectual authority is intended to ensure diversity of values. Diversity of values is thought to improve objectivity because homogeneity of values makes it difficult to recognize assumptions that might be problematic. Equality of intellectual authority allows for the possibility that those who hold different values will be included in the community. When members of the community hold different values, they are more likely to recognize, question, and engage in critical dialogue with those who hold values that they do not share, and in this way contribute to uncovering problematic assumptions.

Brines's revision of the understanding of unequal distribution of domestic labor can be understood as resulting from a more diverse, and consequently more objective, knowing community. Feminists entering economics brought with them an awareness of gender issues that had been absent from traditional analyses. Their different sets of values, interests, and beliefs enabled critical assessment of previous accounts. Longino's vision of the social norms required for objective science is not fully realized in this example, but elements of what make knowledge objective on her account are present.

Kristina Rolin (2017) notes that although Longino's account of objectivity is procedural, it differs from many such accounts in that the procedures that are prescribed are not mechanical or automatic. They require critical discussion among those with differing viewpoints and consequently cannot be reproduced through an algorithm. Together these four norms provide a means for managing bias that can hinder knowledge production. They do this through the identification and critical evaluation of the underlying contextual values that shape the evidence against which the products of science are evaluated. It is worth noting that Longino sees objectivity as a matter of degree, since such norms are often only partially realized.

Longino's version of objectivity is most similar to the conception of objectivity as methodological, involving procedures to ensure objectivity. But it also might be thought of as addressing failures of agent objectivity. The intersubjectivity that is the result of transformational criticism calls for a community of epistemic agents. The process through which knowledge is produced is social and managed through the adherence to a set of norms. The adherence to those norms is what makes knowledge objective to the degree that it is governed by these norms.

> The social value management ideal does not require that non-epistemic values be eliminated from scientific inquiry; instead, it requires that the role of non-epistemic values be analyzed, criticized, and judged as either acceptable or unacceptable by a scientific community which aims to realize the four criteria to a high degree.
>
> *(Rolin 2017: 114)*

While compelling in many ways, Longino's social management model has met with a number of objections since it was first proposed. Among the most worrying is the criticism that the account is too inclusive in allowing all criticism to be considered and responded to. The idea that it is too inclusive has fueled worries that even those who object to the four norms Longino proposes be allowed to participate in critical discussion (Hicks 2011) or that responding to all criticisms would take up an inordinate amount of time pulling scientists away from new research. Rolin has offered several defenses of Longino in this regard (2011, 2017). For the worry about over-inclusiveness, Rolin argues that Longino's four norms would prohibit the participation of those who do not uphold them. As for the concern that responding to criticisms would be too labor intensive and time consuming, she notes that nothing in the norms requires individual scientists to respond to criticism but rather only that the scientific community respond. A division of labor among those in the scientific community could ensure that research would not be stalled.

Kristen Intemann has argued that because Longino's account focuses on diversity of values and not diversity of identity, it does not really address identity-based power differentials within society. While she does not reject Longino's norms, she argues that they need to be supplemented to include identity diversity. She suggests a merging of Longino's approach with features of feminist standpoint theory would do this and that the resulting account would be a feminist standpoint empiricism (Intemann 2010). To better understand this proposal, we now return to feminist standpoint epistemology, first introduced in Chapter 2.

Feminist standpoint theory again

Among Fausto-Sterling's criticisms of the rod and frame test is that the differences in performance might be attributable to experiential differences – that is, how women experience being in a room with a researcher who is a man might be relevantly different from how men experience the same physical circumstance. The awareness of this potential difference comes from having an awareness of the differences in the lived experiences of women and men – that women might be less willing to have the researcher adjust the rod because they might be uncomfortable asserting themselves under these circumstances (or even intimidated when in the dark with a strange man) whereas men are more able to assert themselves or less likely to have had experiences that would make them think that such circumstances are potentially dangerous. Fausto-Sterling's critique raises an alternative hypothesis to explain the results of the research and, in so doing, reveals problematic assumptions that were implicit in the research design.

In Chapter 2, we introduced feminist standpoint approaches and focused on how they might address the question of who knows. Feminist standpoint theory takes the social location of the knower as a crucial factor in knowledge production and identifies differences in interests as arising, in part, from differences in social location. This raises questions for *how* we know objectively in the sense of agent objectivity because it appears to be inconsistent with the idea that objective knowers are "detached" from their own social context and interests. Social location also seems likely to affect, both what we know (ontological objectivity) and pose challenges for how we should go about knowing (methodological objectivity). As a result, feminist standpoint theory would seem to require either rejecting objectivity altogether or proposing a new or revised understanding of objectivity. Most advocates of feminist standpoint theory have opted for a reformulation of objectivity. In this way, they maintain the claim that good science is objective in some sense. We consider two slightly different accounts of objectivity given by standpoint feminists: those offered by Sandra Harding and Alison Wylie.

Sandra Harding: Strong objectivity

For Harding, "The problem was that 'good science' lacked the methodological resources to detect widely held sexist and androcentric assumptions and practices that had shaped these results of research" (Harding 2015: 26). In other words, "good science" understood as operating with traditional standards of objectivity had failed to eliminate illegitimate values from scientific practice and, therefore, results. Harding proposes feminist standpoint approaches and what she calls "strong objectivity" to account for the advances that feminist science has made in detecting these problematic assumptions and practices. Feminist standpoint theory leads to strong objectivity by calling for the scrutiny of the practices through which knowledge is produced. From the dominant frame of reference, the appropriate critical attitude might not be possible as implicit assumptions shared by the scientific community may not be recognized as problematic. Feminist standpoint theory has at its core an "inversion thesis" – the idea that those who are socially, politically, and economically subordinated are, because of their social location, better able to recognize problematic assumptions made by researchers in dominant positions. Standpoint theory proposes that those who are marginalized in these ways may have an epistemic advantage because they are better able to recognize how the dominant power structure shapes research practices and the aims of science. "[A] maximally critical study of scientists and their communities can be done only from the perspective of those whose lives have been marginalized by such communities." (Harding 2004: 136).

Harding thus argues for diversity in knowledge production – specifically a diversity of social location which includes diversity of gender, race, class, sexuality, (dis)ability, geography, and culture. Marginalized knowers are needed for good (strongly objective) knowledge production, because they have the potential to reveal dominant background beliefs and assumptions about methods of research that shape knowledge in support of the status quo and fail to meet the interests and needs of all. In addition, they have the potential to represent their own interests and so may offer alternative understandings of the objects of inquiry. In these ways, those who are members of marginalized groups also may bring epistemic resources that are not available within the dominant framework. Researchers who are marginalized may have an "outsider-within" status (Collins 1986, 2000/2009) since they have access to the lived experience of those studied as well as the dominant culture of those who are doing the studying (the researchers). It is this double-consciousness – the ability to see oneself through the perspective of the dominant culture as well as through one's own lived experience – that aids in recognizing the shortcomings of dominant frameworks.[4]

As we have seen in Chapter 2, epistemic advantage is not automatic and only properly understood when combined with the idea that it is not simply a result of social location or identity, but the conscious awareness of the systemic power dynamics that constitute that location. Because the interests of those who are in dominant social positions differ from those who are marginalized, feminist standpoint approaches posit that those in marginalized positions are better able to understand and theorize the effects of those power relations on the production of knowledge and, as a result, they are also better able to identify problematic values and biases in the practices, processes, and procedures of research that may distort the results.

Sociologists Dorothy Smith and Patricia Hill Collins both use standpoint approaches and their work illustrates the outsider-within concept. Smith notes the mismatch between the lived experience of those who are her subjects of research and the concepts and methods of traditional sociology. She identifies these tensions as "lines of fault" and urges critical feminist sociologists to use such tensions to rethink practices to better capture these experiences. Smith describes this as problematizing everyday life so that what is taken to be "natural" is reassessed and reinterpreted (Smith 1987).

Patricia Hill Collins (1986) makes use of standpoint theory to analyze the effects of race and class. For Collins, neither mainstream sociological accounts nor White feminist approaches fit with the experiences of the women she studies (Collins 2000/2009). Her work provides an example of how standpoint theory is not only compatible with intersectional analysis but actually leads to it. She argues that the researcher who is marginalized may recognize that many of the concepts and procedures adopted by the discipline are problematic when her colleagues do not, precisely because she understands the objects of inquiry both through her disciplinary research tradition and through her own experience as a person occupying a marginalized social location. Through the use of both perspectives, she is able to identify when knowledge produced from the dominant perspective does not fit with the life experiences of those who are marginalized and when it is not adequate to addressing their interests.

Harding sees the work of these scholars (and others) as examples of research that produce better science using standpoint approaches. They are strongly objective rather than merely weakly objective – where weak objectivity means meeting traditional empiricist standards of evidence. Strong objectivity, on the other hand, calls for scrutiny of all phases of research, whereas weak objectivity focuses primarily on phases 3–7. Strong objectivity requires asking whether the subject matter investigated, and the questions asked are appropriate to the goals of research, which will include the liberatory goals of feminism. As Harding puts it, "strong objectivity requires that scientists and their communities be integrated into democracy-advancing projects for scientific and epistemological reasons as well as moral and political ones" (Harding 2004: 136). In this, she embraces the intertwining of political values and science.

An alternative account of objectivity: Alison Wylie

Alison Wylie (2003) offers an alternative account of how feminist standpoint can be understood to produce objectivity. Recall Hacking's claim that "objectivity" is an elevator word. On this understanding, calling some product of science "objective" is a way of valorizing it for having met some other desirable epistemic criteria. Hacking characterized this negatively in that if a claim or theory was objective this indicated that it did not suffer from some epistemic failing such as subjectivity.

Wylie takes objectivity to be a property of knowledge claims and so the question is not whether researchers are objective but whether the products of research should be assessed as such. On her analysis of objectivity, claims are objective to the degree that they conform to some standard set of epistemic virtues. A wide variety of authors, among them Kuhn (1977) and Longino (1990), have proposed such a list. These include empirical adequacy, inferential robustness, explanatory power, internal coherence, and consistency with other established bodies of knowledge (Wylie 2003: 33). We deem claims objective to the degree that they manifest such virtues, whatever the list might be. Where standpoint theory enters the picture is with the acknowledgment that such virtues are rarely, if ever, possible to maximize all at once and so researchers must make choices at each phase of research about which to prioritize for any particular knowledge project. These judgments depend on the interests, purposes, intentions, and goals of the researchers and research community and are consequently sensitive to the social location of the researchers. Feminist standpoint approaches can improve objectivity through weighing the extent to which empirical adequacy, explanatory power, or other virtues are relevant for particular knowledge projects and the degree to which each is relevant.

This account of standpoint theory takes it to be "a purpose-specific epistemic stance, … not a full-service epistemology" (Wylie 2012: 61). In this sense, it is more properly understood as a resource for feminist epistemology and philosophy of science (Crasnow 2009) and as compatible with feminist empiricism (Intemann 2010). As Wylie sees it, standpoint approaches are characterized by three theses: a generic situated knowledge thesis; a systemic situated knowledge thesis; and a thesis of epistemic advantage. The first is best understood as recognition that all knowledge is produced from some socially situated position. There is no "view from nowhere". The second thesis points out that there are epistemic repercussions that result from the systemic nature of social power structures. The thesis of epistemic advantage, as we have already discussed, directs researchers that particular insights may be available to those who are marginalized.

Overall, Wylie's reformulation of standpoint theory calls for researchers to be attentive to the experiential knowledge of those who are marginalized and consider the extent to which those insights are relevant to the goals of their research. This is an ongoing project both because different knowledge projects have different purposes and serve the interests of different groups, but also because the assumptions underlying research are always open to question. Wylie's take on objectivity allows for those interests to determine which of the epistemic virtues to prioritize but does not substitute non-epistemic factors for epistemic virtues. Nonetheless, the judgment about prioritization rests in non-epistemic features of the knowers – features that determine the choice of topics, the questions asked, which observations are relevant, all of which in turn affect choices about data collection and interpretation, and ultimately the conclusions drawn – in other words, all phases of research.

Harding's and Wylie's accounts share a number of features – they agree in identifying the inversion thesis as the core element of standpoint approaches and that it is key to improving objectivity. In earlier work, Harding describes feminist standpoint theory as an alternative feminist epistemology (Harding 1986) – an understanding that Wylie rejects – but Harding's later writings (Harding 2015) describe it more as the sort of resource for feminist science that Wylie sees it as. But Wylie's account more explicitly merges feminist standpoint approaches with feminist empiricism by appealing to standards of good science that empiricists embrace. It is this sort of merging of feminist empiricism with feminist standpoint theory that Kristen Intemann has dubbed "feminist standpoint empiricism" (Intemann 2010).

Feminist holist empiricisms: The pragmatist solution

The shift toward empiricism that we see in standpoint theory combined with changes in empiricism suggests something like a merger of the sort that Intemann proposes. Another way in which contemporary feminist empiricism has developed is that it is strongly influenced by pragmatism. We close out this chapter by considering how to think of objectivity in this context.

Miriam Solomon (2012) identifies several versions of feminist empiricism, differing primarily over the role that values play. She dubs one among these "gap feminist empiricism", which proposes that when faced with theories that are equally empirically successful, values may be used to break the tie.[5] In other words, values are used to fill the gap between the empirical evidence and the theory that results because theory is always underdetermined by evidence. Solomon characterizes Helen Longino's position in *Science as Social Knowledge* (1990) as an example of this form of empiricism.

Another form of feminist empiricism takes values to be subject to the same empirical constraints that other sorts of belief are. In this way, value judgments are objective to the same extent that any judgments are. Solomon refers to this view as "feminist radical empiricism" and identifies Lynn Hankinson Nelson (1990), Sharyn Clough (2003, 2004), and Elizabeth Anderson (2004) as among its advocates. It is radical in that it treats values as empirical and so responsive to evidence. Because they are empirical, values can also serve as evidence. They are both evidence and subject to adjudication by evidence, in the sense that they are part of a holistic belief system that must answer to empirical evidence as a whole – a Quinean web of belief.[6] According to this approach, sexist science fails to be good science because it is less empirically successful than non-sexist science. The success of the theory or belief system serves as an empirical test of the values (and all other claims included in the system). This view is a form of pragmatism.

Solomon also discusses standpoint theory, which she does not consider as an alternative to feminist empiricism but rather sees it as a "methodological tool for feminist empiricists of all kinds" (Solomon 2012: 438). Her discussion also suggests that feminist standpoint theory is better understood as a resource than an epistemological theory, much as we have described Wylie's and Harding's views in the previous section. If we understand it this way, then it serves as a supplement to traditional empiricism – a feminist addition to the procedures, processes, and practices of traditional science.

Solomon ultimately is skeptical that feminist radical empiricism is able to eliminate pernicious values, in part because she doubts that the holism that sustains the view allows for an isolation of any particular problematic value. Audrey Yap (2017) has also considered feminist radical empiricism and expressed similar concerns, although hers are of a more practical nature. She worries that even if values can be adjudicated by evidence, people who hold those values are often not willing to give them up even in the face of such evidence. As a result, Yap worries that this form of feminist pragmatism is likely to fail as a strategy for eliminating sexism and racism in knowledge production.

Conclusions

The examples discussed in this chapter (as well as the last) show that there are potential problems with traditional views about methodological objectivity (as epistemically reliable methods that could be employed by anyone) as well as agent objectivity (as agents who are detached and free of any values). The history of science suggests it is likely impossible for researchers to be wholly "detached" or ensure that scientific decisions are free from values. Yet feminists have also argued that socially situated experiences, interests, and values can benefit knowledge production in certain ways. Thus, the feminist views discussed here reject the VFI of science but have not given up on the possibility of an understanding of "objectivity" that is compatible with that rejection. The accounts we have examined in this chapter shift the focus of agent objectivity so that it is understood as a property of communities rather than individuals. While they disagree on the details, the views discussed aim to identify the features communities should have to enhance knowledge and avoid, or at least minimize, problematic biases. Feminists have also tended to reframe issues of objectivity in terms of methodological objectivity. Objective methods are not (as was traditionally claimed) those that could be employed by anyone with the same results. Rather, they are those that promote the epistemic and social aims of research, including minimizing bias.

With these shifts in how to think about objectivity, feminists also have resources for resolving (or avoiding) the "Bias Paradox" introduced in Chapter 3. Recall that the central concern of the Bias Paradox is that feminists appear to reject certain research as problematic because it involves sexist or androcentric biases, while at the same time endorsing feminist values in research. Feminist approaches to objectivity reject the VFI and thus do not argue that sexist or androcentric values are bad for science because they are *values* as such. Rather, the problem is that such values have gone unnoticed and unanalyzed (perhaps because of a historical lack of diversity within scientific communities, or the extent to which feminists advancing critiques were not granted equality of intellectual authority). Moreover, feminists have argued that such values are either not empirically supported, not widely shared, or, at the very least, do not represent the full range of interests that science ought to address. Thus, the feminist accounts discussed here do not claim that objective science is "value free" but rather that such values need to be identified and scrutinized, which feminist values and approaches can promote.

Where interests are shared and communities are homogeneous what is deemed to be objective is less likely to be disputed. But one aspect of feminist critique of traditional approaches to knowledge production is to highlight that we live in a world where interests are frequently at odds and communities are heterogeneous. This problematizes objectivity in multiple ways that feminist approaches urge us to be attentive to.

Discussion questions

1. Why are the different phases of inquiry relevant for thinking about how gender bias might affect knowledge production?
2. What does the example of the rod and frame test show about research on gender differences?
3. What is inductive risk and how is it relevant to research on gender differences?
4. Why is it important to consider alternative hypotheses in research? What are some problems that arise for doing so that might allow sexist biases to persist?
5. Describe the difference between the ontological and epistemological aspects of objectivity. Which have to do with what we know? Which have to do with how we know?

Additional suggested readings

Clough, S. and W. Loges. (2008). Racist value judgments as objectively false beliefs: A philosophical and social-psychological analysis. *Journal of Social*

Philosophy, 39 (1), pp. 77–95. An argument that sexist and racist values are subject to empirical adjudication and can be shown to be false (relevant to the discussion of holism in this chapter).

Hamington, M. and C. Bardwell-Jones (eds.). (2012). *Contemporary feminist pragmatism*, New York: Routledge, 2012. A deeper exploration of feminist pragmatism.

Rooney, P. (2017). The borderlands between epistemic and non-epistemic values. In: K. C. Elliott and D. Steel (eds.). *Current controversies in values and science*. New York: Routledge, pp. 31–45. A discussion of the difficulties in drawing a sharp distinction between epistemic and non-epistemic values.

Notes

1 We use "phases" here rather than "stages" to avoid the implication that these occur in chronological order. Anderson (2004) and Brown (2020) have similar accounts although our breakdown differs from each of theirs slightly. Anderson uses "stages", whereas Brown uses "phases" and emphasizes that each is a point at which decisions that affect the research are made, much as we are doing.
2 What we have in mind is the way that methods (like using quantitative or measuring methods) appear to provide objectivity but at the price of having already made choices about what is relevant (what is measurable and how it is to be measured) and so affect what we can know. See Montuschi's (2021) discussion of Porter (1995). Montuschi argues that in considering objectivity we have to ask whether what is taken to be relevant is adequate to achieve the goals of research.
3 In much of this literature, researchers use the language "male" and "female" and indeed they take this work to support that there are essential biological differences between two distinct kinds: males and females. Much of the feminist work related to this literature on sex differences is also aimed at challenging this binary and essentialist view.
4 The notion of "double-conscious" is due to W. B. E. DuBois (1903/1993).
5 Solomon cites Intemann 2005 as first describing this view as gap feminism.
6 Although Clough's version relies on the work of Donald Davidson, whose holism is based on Quine's. She explains how the account is developed from Davidson's work in Clough 2003.

References

Abdul Razzak, R., and J. Bagust. (2022). Perceptual lateralization on the rod-and-frame test in young and older adults. *Applied neuropsychology: Adult*, pp. 1–7.
Anderson, E. (2004). Uses of values judgments in science: A general argument, with lessons from a case study of feminist research on divorce. *Hypatia: A Journal of Feminist Philosophy*, 19, pp. 1–24.
Brines, J. (1994). Economic dependency and the division of labor. *American Journal of Sociology*, 100 (5), pp. 652–688.
Brown, M. J. (2020). *Science and the moral imagination: A new ideal for values in science*. Pittsburgh, PA: University of Pittsburgh Press.
Clough, S. (2003). *Beyond epistemology: A pragmatist approach to feminist science studies*. Lanham, MD: Rowman and Littlefield.

Clough, S. (2004). Having it all: Naturalized normativity in feminist science studies. *Hypatia: A Journal of Feminist Philosophy*, 19 (1), pp. 102–118.

Collins, P. H. (1986). Learning from the outsider within: The sociological significance of Black feminist thought. *Social Problems*, 33 (Special theory issue), pp. S14–S32.

Collins, P. H. (2000/2009). *Black feminist thought*. New York: Routledge.

Crasnow, S. (2009). Is standpoint theory a resource for feminist epistemology? *Hypatia: A Journal of Feminist Philosophy*, 24 (4), pp. 189–192.

Douglas, H. E. (2009). *Science, policy, and the value-free ideal*. Pittsburgh: Pittsburgh University Press.

DuBois, W. E. B. (1993/1903). *The souls of black folk*. New York: Alfred A. Knopf Publishing.

Elliott, K. C., and D. J. McKaughan. (2009). How values in scientific discovery and pursuit alter theory appraisal. *Philosophy of Science*, 76 (5), pp. 598–611.

Fausto-Sterling, A. (1985). *Myths of Gender*. New York: Basic Books.

Fausto-Sterling, A. (1993). The five sexes. *The Sciences*, 33 (2), pp. 20–24.

Gould, S. J. (1996). *Mismeasure of man*. New York, NY: WW Norton & Company.

Harding, S. (1986). *The science question in feminism*. Ithaca, NY: Cornell University Press.

Harding, S. (2004). Rethinking standpoint epistemology: What is 'strong objectivity?' In: S. Harding (ed.), *The feminist standpoint reader*. New York: Routledge, pp. 127–140.

Harding, S. (2015). *Objectivity and diversity: Another logic of scientific research*. Chicago: University of Chicago Press.

Hicks, D. (2011). Is Longino's conception of objectivity feminist? *Hypatia: A Journal of Feminist Philosophy*, 26 (2), pp. 333–351.

Intemann, K. (2005). Feminism, underdetermination, and values in science. *Philosophy of Science*, 72 (5), pp. 1001–1012.

Intemann, K. (2010). 25 years of feminist empiricism and standpoint theory: Where are we now? *Hypatia: A Journal of Feminist Philosophy*, 25, pp. 778–796.

Kitcher, P. (2001). *Science, truth, and democracy*. New York: Oxford University Press.

Kourany, J. A. (2010). *Philosophy of science after feminism*. Oxford: Oxford University Press.

Kuhn, T. S. (1977). *The essential tension: Selected studies in scientific tradition and change*. Chicago: University of Chicago Press.

Longino, H. (1990). *Science as social knowledge: Values and objectivity in scientific inquiry*. Princeton, NJ: Princeton University Press.

Longino, H. (1993). *Subjects, power, and knowledge: Description and prescription in feminist philosophies of science*. In: L. Alcoff and E. Potter (eds.), *Feminist epistemologies*, pp. 101–120.

Longino, H. (2002). *The fate of knowledge*. Princeton, NJ: Princeton University Press.

Maccoby, E. E., and Jacklin, C. N. (1974). Myth, reality and shades of gray: What we know and don't know about sex differences. *Psychology Today*, 8, pp. 109–112.

Maney, D. L. (2016). Perils and pitfalls of reporting sex differences. *Philosophical transactions of the Royal Society of London. Series B: Biological sciences*, 371 (1688), 20150119.

Montuschi, E. (2021). Finding a context for objectivity. *Synthese*, 199, pp. 4061–4076.

Nelson, L. H. (1990). *Who knows: From Quine to a feminist empiricism*. Philadelphia: Temple University Press.

New York Times. (1912, May 10). Sex differences in brain. https://timesmachine.nytim
  es.com/timesmachine/1912/05/10/104896143.html?pageNumber=4.
Okruhlik, K. (1994). Gender and the biological sciences. *Canadian Journal of Philoso-
  phy*, 24(sup1), pp. 21–42.
Porter, T. (1995). *Trust in numbers: The pursuit of objectivity in science and public
  life*. Princeton, NJ: Princeton University Press.
Richardson, J. T. (1994). Gender differences in mental rotation. *Perceptual and Motor
  Skills*, 78(2), pp. 435–448.
Rolin, K. (2011). Contextualism in feminist epistemology and philosophy of science. In:
  H. Grasswick (ed.), *Feminist epistemology and philosophy of science: Power in
  knowledge*. Dordrecht: Springer, pp. 25–44.
Rolin, K. (2017). Can social diversity be best incorporated into science by adopting the
  social value management ideal? In: K. C. Elliott and D. Steel (eds.), *Current con-
  troversies in values and science*. New York: Routledge, pp. 113–129.
Rudner, R. (1953). The scientist *qua* scientist makes value judgments. *Philosophy of
  Science*, 20, pp. 1–6.
Smith, D. (1987). *The everyday world as problematic*. Boston: Northeastern University
  Press.
Solomon, M. (2012). The web of valief: An assessment of feminist radical empiricism.
  In: S. L. Crasnow and A. M. Superson (eds.), *Out from the shadows: Analytical
  feminist contributions to traditional philosophy*. Oxford: Oxford University Press,
  pp. 435–450.
Witkin, H. A. & S. E. Asch. (1948). Studies in space orientation: III. Perception of the
  upright in the absence of a visual field. *Journal of Experimental Psychology*, 38, pp.
  603–614.
Wylie, A. (2003). Why standpoint matters. In: R. Figueroa and S. Harding (eds.), *Sci-
  ence and other cultures: Issues in philosophy of science and technology*. New York:
  Routledge, pp. 26–48.
Wylie, A. (2012). Feminist philosophy of science: standpoint matters. *Proceedings and
  Addresses of the American Philosophical Association*, 86 (2), pp. 47–76.
Yap, A. (2017). Feminist radical empiricism, values, and evidence. *Hypatia: A Journal
  of Feminist Philosophy*, 31 (1), pp. 58–73.
Yuan, L., F. Kong, Y. Luo, S. Zeng, J. Lan, and X. You. (2019). Gender differences in
  large-scale and small-scale spatial ability: A systematic review based on behavioral
  and neuroimaging research. *Frontiers in Behavioral Neuroscience*, 13, p. 128.

5

WHAT DON'T WE KNOW?

Overview

The view of science as uniformly progressive has been widespread in philosophy of science. The result is the failure to examine the ways in which commitments to particular research programs create strategic lacunae in our knowledge that benefit some and perpetuate the epistemic disadvantages of others. The aim of this chapter is to introduce readers to the field of agnotology – the study of how ignorance is produced and maintained – and look at it specifically in the context of philosophy of science. We examine how feminist work on epistemologies of ignorance has contributed to our understanding of what ignorance is, how it occurs, and when it is problematic. We distinguish several different types of ignorance and identify a variety of ways in which they are produced or maintained (both actively and passively).

Introduction

In 2006, activist Tarana Burke began the MeToo movement on social media, encouraging people to break their silence and publicize their experiences of sexual assault and harassment. The movement gained considerable steam in 2017, when actress Alyssa Milano urged people to use #MeToo, in the midst of widespread sexual abuse accusations against film producer (and now convicted sex offender) Harvey Weinstein. Part of what was so powerful about the MeToo movement, was that it made visible something that had been previously unknown to many: the incredibly high prevalence of sexual assault and harassment, especially in the workplace. Not only did the movement increase social awareness of sex crimes, it also empowered some to perceive

DOI: 10.4324/9781032693781-5

their own experience in a new way. It made it possible for survivors to see their experiences as fitting into a pattern of behaviors driven by unequal power relationships. Some women who had sexual encounters with employers, such as Monica Lewinsky (former White House intern under President Bill Clinton) and a variety of NBC News employees who worked with NBC anchor Matt Lauer, had previously described encounters as "consensual", but later acknowledged that the consent came partly or largely from fear of what would happen to their careers if they refused. Thus, it made visible assumptions about consent and the existence of certain power dynamics and provided new conceptual tools to survivors.

While the MeToo movement made certain knowledge visible, the use of the MeToo hashtag also obscured certain aspects of sexual assault and harassment. The surge of its use, particularly among wealthy White cisgendered celebrities, made less visible the very thing that Tarana Burke's original movement was trying to call attention to: the fact that women of color, Indigenous women, transwomen, as well as other vulnerable women are disproportionately targeted by gender violence. In addition, the impact of MeToo (as well as some of the reaction to it) suggests that there was still a lack of knowledge about the inequalities and power dynamics that exist in the workplace and elsewhere in society. That it seemed to be *new knowledge* for so many is perplexing, given that instances of sexual violence and laws against it had been around for decades. How is it that certain facts (such as the prevalence or disproportional impacts of sexual assault and harassment) can remain unknown, or be obscured, for large populations and certain individuals?

Traditional epistemology focuses on the question: How is *knowledge* produced and achieved? Yet there are important insights to be gained by flipping that question to: How is *ignorance* produced and maintained? This is the central question of the field of agnotology – or the study of ignorance – that gained traction in the 1990s (Proctor 1995) and was also developed by feminists, critical race theorists, and decolonial theorists working on "epistemologies of ignorance" (Frye 1983; Mills 1997; Tuana 2004; Sullivan and Tuana 2007). As Shannon Sullivan and Nancy Tuana explain, "The epistemology of ignorance is an examination of the complex phenomena of ignorance, which has as its aim identifying different forms of ignorance, examining how they are produced and sustained, and what role they play in knowledge practices" (Sullivan and Tuana 2007: 1).

The central focus of this chapter will be to examine how feminist scholars have furthered our understanding of ignorance and its importance as an epistemically important – as well as politically significant – phenomenon. In previous chapters, we have already seen how ignorance can be created about scientific phenomena in certain ways. For example, it may be that certain research questions are neglected. In some cases, the ways in which something is studied (e.g., heart disease) can systematically neglect certain groups (such as women or people of color). This chapter will look more deeply at what ignorance is and how it can be produced and maintained.

What is ignorance?

The concept of ignorance is often thought to have a negative connotation (and most people would probably object to being called "ignorant"). Yet, ignorance is not always bad and in some cases it may even be beneficial. For example, we utilize ignorance to counteract bias when we blind clinical trials. Knowledge of whether one received a placebo or an experimental drug might influence the results of an experiment and even undermine the interests of patients. So, this information is often deliberately withheld from both patients and researchers. There is also knowledge that we do not seek because it would be unethical to do so. We do not force human research subjects into randomized clinical trials to determine whether certain surgical procedures are better than current treatments because this would violate patients' rights to autonomy in making their own treatment decisions. Yet, as a result, there is some knowledge or some evidence that we agree is unethical to acquire. In addition, certain kinds of knowledge may be particularly dangerous or thwart our security interests, such as how to create a biological weapon, so that it is best to restrict access to that knowledge. In our day to day lives, there may be some knowledge that we would prefer not to know, such as when someone says something unkind about us. But ignorance can clearly also be problematic, particularly because knowledge is often necessary for guiding individual decisions or actions and justifying policies. While we will not provide a full account here of when or under what conditions ignorance is bad or wrong, this is an important topic (see for example, Fernández Pinto 2015, 2020). For our purposes, we only want to recognize that "ignorance" is a phenomenon that can be useful or beneficial, even though it can also be problematic. Our focus here will be on feminist and critical race contributions to these discussions, which have been particularly concerned with the ways in which ignorance is used to maintain, deepen, or reinforce systemic inequalities (e.g., Mills 1997) even though some have pointed out that ignorance can also be a coping mechanism for those subjected to systemic oppression (Sullivan and Tuana 2007).

Some have equated ignorance with the absence of knowledge (Le Morvan 2010), a lack of true beliefs (Peels 2011), or uncertainty (Smithson 1989). Yet many have argued that ignorance should not be equated with merely a lack of knowledge or a cognitive deficit (Sullivan and Tuana 2007; Proctor 2008: 3–6). There are many things that we lack knowledge about for reasons discussed in earlier chapters. We cannot know everything given our practical limitations as humans and the constraints on our time. Social theories of ignorance, however, focus on the ways in which social and cultural forces construct and maintain ignorance. They examine how such forces work to bring about not only a lack of knowledge, but make it difficult, if not impossible to know certain things (Smithson 2008: 214–215). Thus, the focus here will be on ignorance not as a "neglectful" epistemic practice, but as "a substantive epistemic practice in itself" (Alcoff 2007: 39). Feminist philosopher Marilyn Frye (1983), for example, called attention to the ways in which White ignorance is an active force (even among White feminists who consider themselves anti-racist), cultivated and supported by social institutions and structures that enable White people to be oblivious to White privilege. Frye explains that:

> ignorance is not something simple: it is not a simple lack, absence or emptiness, and it is not a passive state. Ignorance of this sort—the determined ignorance most white Americans have of American Indian tribes and clans, the ostrich-like ignorance most white Americans have of the histories of Asian peoples in this country, the impoverishing ignorance most white Americans have of Black language—ignorance of these sorts is a complex result of many acts and many negligences.
>
> *(Frye 1983: 118)*

Such forces and practices can operate in ways that are either intentional or unintentional. In science, we ask and pursue certain research questions over others, focusing our attention on some problems or some aspects of problems and not others. This creates gaps about things we don't know, but it can be the product of inattention, rather than a conscious choice to ignore or disregard other areas of investigation. Thus, ignorance can be a consequence of the selective nature of inquiry (Proctor 2008: 7). Of course, some inattention can also be willful or neglectful as Frye points out, even if it is not consciously intentional. Some ignorance, moreover, may be deliberately created. Several scholars have focused on the ways in which industries and private interests aimed to manufacture doubt and confusion about scientific evidence on the health consequences of smoking, climate change, or the risks posed by chemicals to stall or prevent regulations (Proctor 2011; Oreskes and Conway 2010; Michaels 2020).

One reason that feminists began to focus on epistemologies of ignorance grew from a desire to call attention to the ways in which ignorance has been produced and maintained about women and other intersecting social groups (Tuana 2004, 2006; Kourany 2015). Feminists, critical race theorists, critical disability theorists, and decolonial scholars also wanted to identify and challenge the ways in which ignorance reinforces or perpetuates inequalities or injustice, particularly along intersecting lines of gender, ethnicity, ability, and class (Harding 2008; Medina 2013, 2017; Gilson 2011; Tremain 2017).

Varieties of ignorance

Nancy Tuana (2006) uses the women's health movement of the 1970s and 1980s to suggest a taxonomy of the types of ignorance that were revealed in relation to women's health. Tuana characterizes the women's health movement not only as a political movement aimed at empowering women with knowledge about their bodies and health, but also as a movement of "epistemological resistance" geared at "undermining the production of ignorance about women's health and women's bodies in order to critique and extricate women from oppressive systems often based on ignorance" (2006: 2). Using examples from this movement, Tuana's helpful taxonomy of ignorance (2006: 4–15) includes:

1. What we don't know, but don't care to know.
2. What we don't know that we don't know.
3. What they don't want us to know.
4. What they don't know and don't want to know.
5. What we cannot know.

First, there are things we do not know, but don't care to. As noted in Chapter 3, knowledge, including the pursuit of scientific knowledge is linked to *what we are interested in knowing*. We are not just interested in true beliefs, but facts about the world that are particularly significant, interesting, or useful in some way. For example, since the 1960s, hormone-based contraceptives were directed toward and tested on women. In early development of oral contraceptives, it was widely believed that men would be less motivated than women to use contraceptives and less willing to accept side effects (Oudshoorn 2003). The assumption was that men would never accept, for instance, the risk of decreased libido associated with hormone contraceptives. In contrast, no one seemed to worry about decreased libido in women, given the pervasive stereotype that women did not have much sexual desire to begin with (Arditti 1977: 123; see also Tuana 2006: 4–5). Moreover, pharmaceutical companies assumed that preventing conception was women's responsibility (not men's) and thus they just did not see it as a marketable investment. Tuana's point here is that we still do not know about safe or effective hormonal contraceptives for men, because the pharmaceutical companies were not interested in investigating something they thought would not be profitable (Tuana 2006: 5). It was not that this possibility was *overlooked* – indeed, it was actively considered as an area of investigation and rejected. The decision not to know was the product of the power and privilege to do the research, as well as sexist stereotypes. This kind of ignorance is very much the product of the ways that values can shape what is studied and how research questions are framed, as discussed in Chapters 3 and 4.

A second type of ignorance involves things we don't even know that we don't know because our current interests or state of knowledge block such knowledge or make it invisible. This type of ignorance is difficult to identify and, when we identify it in hindsight, it seems similar to the first category. Tuana uses the example of the lack of knowledge about the anatomy of the female clitoris that occurred in part because the scientific understanding of the anatomy and genitalia focused on knowledge about reproduction (in which the clitoris was not viewed as having any role). The focus on reproduction – as opposed to sexual pleasure – rendered the clitoris, which was assumed to be irrelevant to reproduction, as irrelevant to investigation. Researchers did not know what they did not know (they did not know that they were ignorant about this) because they simply did not view it as a discrete or significant object of study (Tuana 2006: 7). This is distinct from the first kind of ignorance because it was not an area of inquiry actively considered and rejected as unworthy – rather it was not viewed as an object of inquiry at all.

Third, Tuana identifies cases of ignorance that are created deliberately and systematically for certain groups: what they don't want us to know. She provides the example of the pharmaceutical industry's concealment of known risks and dangerous side effects of estrogen in female contraceptives because they were trying to protect their profit margins (Tuana 2006: 9). Indeed, there are many cases where private industry withheld, hid, or suppressed information to avoid regulation, litigation, or loss of profits (e.g., Michaels 2020). Yet, while this type of ignorance is often assumed to be problematic, "what they don't want us to know" may also include things such as instances of "dangerous knowledge" that are restricted, classified, or suppressed because widespread access may have incredibly harmful consequences (see Proctor 2008: 2–3).

Fourth, Tuana discusses cases of *willful ignorance*, or cases of what we don't know and don't want to know. White ignorance about systemic racism might be best understood as an example of willful ignorance. As mentioned earlier, Frye described ignorance about White privilege as willful – a determined ignorance that is the result of many acts of negligence. Willful ignorance is distinct from those types of ignorance previously discussed because it involves employing active strategies to *avoid* knowing, including neglecting evidence that might force us to have knowledge of something we would prefer not to know. Charles Mills (1997) has pointed out that it is not just that some White people are ignorant about their own privilege, but that they refuse to understand the oppressive conditions experienced by non-White people, as well as the institutions, beliefs, and practices that underlie those inequalities. When a society is structured by inequality and relations of power and domination, the dominant class has interests in viewing the world in particular ways, as a place where inequalities are justified or do not exist

(Mills 1997). People with privilege have an interest in believing that the world is fair and just and that they have what they have as the result of hard work and virtuous behavior, such that what they have (and what they avoid) is deserved. They do not have an interest in learning how the world is not that way because it would have potentially negative implications. Thus, as James Baldwin argued, "White America remains unable to believe that Black America's grievances are real; they are unable to believe this because they cannot face what this fact says about themselves and their country" (Baldwin 1985: 536).

Elizabeth Spelman (2007) provides a powerful analysis of how we might understand Baldwin's claim. It is not *merely* that White people fail to be interested in learning about injustices against Black America or that the experiences of White people block their achievement of certain knowledge. It is also that when confronted with testimony or evidence about injustices against Black people many White people *immunize* themselves against any criticism or evidence that would correct their misunderstandings (Spelman 2007: 119). Thus, even when presented with evidence, for example of police brutality against Black people, there is still often a stubborn refusal to accept that systemic racial injustice in policing exists. Evidence of police brutality is discredited (surely the suspect had to have been doing something wrong and so their interpretation is not to be believed) or dismissed (the use of force was reasonable because the suspect did not "comply" or it was reasonable for police to believe they were in danger). Indeed, it is often only in exceptional cases where extreme acts are committed and there is inescapable evidence – such as multiple videos of the murder of George Floyd by police in Minneapolis in 2020 – that there is widespread recognition among White people that some wrong has been committed. Even in such cases, they are often taken to be exceptional – that is, there is still resistance to believing that this is evidence for widespread or systemic racism.

Willful ignorance occurs when we don't want to believe something. Yet, according to Spelman, it is also distinct from self-deception. Self-deception occurs when we convince ourselves that something is the case because we want to believe it to be the case. With respect to systemic racism, many try to avoid considering the issue or drawing any conclusions altogether. We *fear* that it might be true, we have an interest in not believing it, and so we avoid thinking about it as much as possible or construct understandings that allow us to reject or minimize whatever evidence might compel us to believe (Spelman 2007: 121). White ignorance is thus an instance of willful ignorance insofar as it involves what Tuana describes as an "active ignoring of the oppression of others and one's role in that exploitation" (Tuana 2006: 11).

White ignorance is not the only instance of willful ignorance. As mentioned at the beginning of this chapter, the MeToo movement gained momentum because of the multiple charges against film producer Harvey

Weinstein. Yet, at the very same time that many in Hollywood decried what Weinstein and other powerful figures had done and proclaimed that survivors were to be believed, many were curiously quiet or even defended actor and director Woody Allen.

For over 20 years, Dylan Farrow has consistently accused Woody Allen (her adoptive father) of sexual assault and abusing her in 1992 when she was 7 years old. Allen has denied the allegations against him (like Harvey Weinstein) and suggested that Dylan was convinced the abuse had occurred because she had by coached by her mother Mia Farrow. Allen claims that Mia Farrow was motivated by anger because Allen admitted to "an affair" and subsequently married a separate adopted daughter of hers. As Dylan Farrow points out in an OpEd in 2017, this was not a "he said, child said" situation (Farrow 2017). Multiple friends and family members witnessed a pattern of inappropriate sexual touching and behavior by Allen toward Dylan for a significant period of time, which led him to agree to go into therapy (which he was in when the assault occurred). Three eyewitnesses corroborated Dylan's account, experts have said that there no signs that Dylan's memory was the result of suggestion and coercion, and her behaviors and descriptions fit a pattern typically seen in credible abuse cases. A judge subsequently denied Allen custody of Dylan because of the risk he posed, and a prosecutor found there was "probable cause" to believe the allegations were true but declined to charge Allen out of concern for the challenges a traumatized minor victim might face at trial. Yet Allen was not rejected by Hollywood like Weinstein. Somehow, he was still celebrated and given awards for his creative work.

Most importantly, for our purposes, women and men in Hollywood who worked with Allen vociferously supported MeToo and rejected Weinstein while managing to evade having to address charges against Allen. In 2017, actress Kate Winslet gave an interview to *Variety* where she stated of the accusations about Weinstein: "The fact that these women are starting to speak out about the gross misconduct of one of our most important and well-regarded film producers, is incredibly brave and has been deeply shocking to hear" (Setoodeh 2017). Yet that same year she starred in Allen's film *Wonder Wheel* and in an interview with the *New York Times,* when asked whether she was uncomfortable working with Allen given the allegations against him she said,

> I didn't know Woody and I don't know anything about that family. As the actor in the film, you just have to step away and say, I don't know anything, really, and whether any of it is true or false. Having thought it all through, you put it to one side and just work with the person. Woody Allen is an incredible director.
>
> *(Ryzik 2017)*

Similarly, in an interview with the *Hollywood Reporter,* actress Blake Lively said of Weinstein:

> It's important that women are furious right now. It's important that there is an uprising... The number one thing that can happen is that people who share their stories, people have to listen to them and trust them, and people have to take it seriously.
>
> *(Gardner 2017)*

But on the subject of Allen, she said, "It's very dangerous to factor in things you don't know anything about. I could [only] know my experience. And my experience with Woody is he's empowering to women" (Zeitchik 2016). These statements also seem to reveal a pattern of willful ignorance – actively avoiding whether Dylan Farrow was sexually abused and assaulted by Woody Allen because of a fear that it might be true.

Notice that neither of these actresses (and there were many others who said similar things) is prepared to say the allegations against Allen are *false,* presumably because that would require them to consider the testimony of Dylan Farrow, testimony that in other contexts they believe should be "listened to" and "trusted". But they do not have an interest in believing the claims against Allen are true, perhaps because they fear this would have prevented an important career opportunity or made them culpable for working with him. Instead, it is easier for them to "not know" about it, actively ignoring the evidence and avoiding any conclusions one way or the other. These examples fit into a larger historical tendency that Tuana points out, which is willful ignorance about incest, despite large numbers of children who are incestuously molested each year in the United States (Tuana 2006: 12–13).

While the first four types of ignorance that Tuana identifies are largely thought to be problematic types of ignorance, the final type of ignorance on Tuana's taxonomy is "what we cannot know". There are many things we cannot know in virtue of our limitations as humans. For example, there may be physical limitations on what we can perceive about the physical world. There are practical limits on what we can know, given our time and limited resources. There are cognitive limitations that, for instance, might preclude us from having or understanding the experiences of others. Maria Lugones (2003) and Maria Ortega (2006) have used the term "loving ignorance" as involving an acceptance of our basic cognitive limitations in understanding the experiences of others. Despite whatever good intentions we might have, there are limitations on what we can know because there are certain experiences that cannot fully be shared. Loving ignorance can occur, however, when we recognize our own limitations. For Lugones, it requires a loving perception for those who have been historically marginalized. It requires not ignoring them, dismissing them, or stereotyping them, but also realizing that although much of these experiences can be comprehended or understood, there are some that cannot (Lugones 2003: 83). It involves a loving respect for the limits of our own knowledge and experiences. According to Ortega, achieving loving ignorance requires more than merely trying to understand the experiences of others. In fact, she argues that many White feminists believe it is sufficient to read about the experiences of non-White feminists, but that this too can give rise to a sort of arrogance or confidence that one fully understands the lives and experiences of women of color. For Ortega, achieving loving ignorance requires active engagement with women of color, to build relationships and learn about their experiences (Ortega 2006: 67).

This taxonomy is not intended to be exhaustive and there are some forms of ignorance, such as White ignorance, that could involve different categories in this taxonomy. Some instances of White ignorance might be understood as not caring to know, while in some cases, current knowledge and interests block knowledge about inequalities. In some cases, we may think of the ignorance as willful or negligent, and, in some cases, even those who resist White supremacy and aim to be anti-racist may still be "lovingly" ignorant because they can never fully know "what it's like" to live in a racially unjust society as a Black person. This taxonomy is helpful, however, in getting a sense of the range of types of ignorance that feminists and critical race theorists have identified.

How ignorance is produced and maintained

Cutting across the different types of ignorance that have been identified, there are several strategies and tactics that are implicitly or explicitly used to produce or maintain ignorance. This section will aim to provide examples of some of these strategies.

Neglect

As we have seen, one of the ways that ignorance can be produced and maintained is by neglecting certain areas of study, research questions, experiences, or phenomena. Sometimes this neglect is unintentional, as in cases of "not knowing what we don't know". Certain phenomena may be neglected because we aren't motivated to investigate them (e.g., male contraceptives) or because our current theories and interests block viewing them as an object worthy of investigation (e.g., knowledge of the clitoris). Adriene Mayor (2011) argues that ancient Mediterranean knowledge about fossils was neglected in the history of science because scientists and academics tended to ignore non-elite sources and viewed fossil accounts as folklore or fiction that was not to be taken seriously. As a consequence, modern paleontologists were ignorant about the earliest fossil discoveries.

Neglect can also happen in ways that are more active – or involve some varying degree of intentionality. As we saw in cases of willful ignorance, we can choose to ignore or refuse to consider some area of inquiry. Media companies often choose not to cover certain stories because some stories either do not align with their values or with the interests of their target consumers. This is one reason that people who receive their news from different sources sometimes have different beliefs – it is because what is being covered (and not covered) is potentially different.

Similarly, consider debates over which aspects of US history to teach or not teach in public schools. Given limited time and resources, not every detail about everything that happened in the history of the US should be taught. While there is likely agreement that some aspects of history will be neglected, there is disagreement about what *ought* to be neglected and what ought to be taught. Critical race theorists, broadly speaking, believe that it is important to use race as a category of analysis to examine how institutions, laws, policies, and practices may impact racial groups differently and/or may involve racial inequalities, stereotypes, or racist norms. While critical race theory (CRT) is an academic theory not taught in K-12 classrooms, it might be used to argue that the teaching of history ought to include attention to race – for example the institution of slavery – or how race shaped certain aspects of the world (such as our legal, political, or healthcare systems). Some politicians, however, argue that this focus mostly portrays US history in a negative light that may lead students to believe that the US is an inherently bad country. Proponents of CRT argue that if we do *not* use race as a category of analysis in thinking about the world, failing to consider how, for example, the institution of slavery, Jim Crow segregation laws, and voting laws impact groups differently along racial lines, generate inequities, or are enforced differently, we will be neglecting important salient facts about the world. To neglect this history (and how it shapes our current institutions) produces and maintains ignorance about inequalities.

Of course, neglect can occur across the political spectrum. In 2016, the Democratic Party made a strategic decision to neglect voters in rural areas and focus on winning voters in urban and suburban areas (Bottomiller Evich 2016). The result was not just that rural voters overwhelmingly voted for Republicans, it also meant that Democratic Party officials, as well as democratic politicians were (and arguably remain) systematically ignorant about the circumstances, needs, and values, of rural Americans. Rural areas often face greater challenges in generating jobs and opportunities, particularly as international trade reduced US manufacturing, family-owned farms were pushed out by agribusiness, and mining jobs were reduced as a result of technological developments and concerns about climate change (Wuthnow

2018) and recent job growth has disproportionately benefited urban areas. Many Democratic policy proposals fail to be informed by this knowledge, potentially due to neglect.

Suppression

Ignorance can also be produced and maintained by suppressing knowledge. Suppression occurs when there is a failure to disseminate or take up knowledge, such that it becomes lost or forgotten. Like neglect, it can occur in passive and active ways and be done with either good or bad intentions. Suppression can occur through intentional denial of access to certain kinds of knowledge, ideas, or evidence. For example, restrictions that were adopted in the early 2020s to prohibit or restrict the teaching of certain topics or using books that might touch on those topics, such as racism or anything having to do with sexuality. Suppression can occur through well-intentioned deliberate secrecy by, for example, classifying knowledge that might be too "dangerous" to be publicly known. Suppression can occur in science, when scientists fail to share data or be completely transparent about their methodologies. This can happen in cases of industry-funded science where there are proprietary issues or in cases of publicly-funded science where national security is at stake (or claimed to be at stake). Knowledge can also be deliberately suppressed by marginalized groups. As Adriene Mayor (2008) discusses, some Native American groups decided to withhold certain fossil knowledge from outsiders, and even limited its dissemination within their nation, for fear that certain knowledge might be harmful or used against them. For example, many Navajos avoid touching or talking about anything to do with death, including dinosaur fossils on their land (Mayor 2008: 176).

Failures to disseminate knowledge can also occur in virtue of structures or forces that have the effect of making the knowledge inaccessible. For instance, differences in language or in the means used to disseminate information might make that information less comprehensible or visible to some. Sandra Harding and Brenny Mendoza have argued that Latin American feminist philosophy, while generating important work about the history of science and innovative contributions to philosophy, has largely remained invisible among Anglo scholars who do not speak Spanish or Portuguese (Harding and Mendoza 2021). Knowledge produced by Indigenous women with long traditions for disseminating knowledge through oral histories and storytelling has been lost both because of language barriers but also because oral histories are sometimes excluded as not "reliable or objective".

Suppression can also happen through a failure to cite or credit knowledge to others, or to show how it builds on what was previously known. Adriene Mayor (2008) reveals that landmark discoveries of some of the first American fossils were made by Indigenous peoples whose work was usually not credited. French naturalist George Cuvier was one of a few Anglo researchers who did try to disseminate and credit Indigenous fossil knowledge. Cuvier noted, for instance that natives of Canada and the Ohio Valley identified mastodons as the "grandfather or ancestor of the buffalo" and wrote about Iroquois and Delaware accounts of mastodon fossils in the Ohio Valley. Yet, when his work was translated into English nearly two decades later, Cuvier's extensive section on Native American accounts was neither translated nor commented on (Mayor 2008: 172). This created ignorance about Indigenous fossil knowledge and also had the distorting effect of making it seem as though fossil discoveries in North America were "European", reinforcing the colonial myth that no one existed or had knowledge there before it was "discovered" by Europeans.

Discrediting

In addition to failing to give proper credit to knowledge, ignorance can also be produced and maintained by actively "discrediting" bodies of knowledge or knowers. Indigenous and local knowledge has often been discredited as "myth" or "folklore" in contrast to scientific knowledge. Because such knowledge was not seen as hypothesis driven and recorded in writing the same way as Western science, it was viewed as unscientific and not rigorously justified, therefore not knowledge. There are, however, many examples of reliable and empirically adequate knowledge produced by Indigenous knowledge practices. Micronesian islanders developed systems for navigation (Harding 2015). Extensive Indigenous knowledge of plants and their medicinal properties were exploited throughout the Atlantic world (Schiebinger 2017). Even today, major "discoveries" in conservation biology and ecology are often found to have long been known through traditional ecological knowledge (Nicholas 2018). Yet, by categorizing these bodies of knowledge as "unscientific" and therefore not "real knowledge", ignorance was produced about the domain of knowledge to which they pertained (e.g., navigation, oceanography, botany, conservation biology), but also about the history of science, as well as about the contributions and capabilities of certain groups. This is part of a larger tendency of what we might call the "scientification" of knowledge, or the assumption that only the products of Western science can be called *knowledge*. It builds on the assumption, mentioned in Chapter 2, that knowledge is primarily *propositional* and that it is only justified through repeated empirical testing of those propositions. Yet there may be other ways of knowing, such as experiential knowing, for example, knowledge that is gained through practice, rather than by studying particular propositions in a systematic and controlled way. This is the sort of experiential knowledge that midwives have about prenatal care, childbirth, and the wellbeing of mothers. Such experiential knowledge was dismissed as "non-scientific" and "nonskilled" as physicians began to professionalize obstetrics and gynecology in the 19th century (Dalmiya and Alcoff 1993). The knowledge of midwives was discounted and discredited as unscientific, uneducated, and lacking expertise, such that midwives began to face requirements for licensing and restrictions on practicing, many of which still exist today. The assumption here is that knowledge can only be generated by a particular sort of formal training and education, to which many women lacked access.

Another way of discrediting knowledge is to undermine it as uncertain – a strategy which has been employed against academic scientists themselves. During the 1960s, the Tobacco industry sought to discredit science showing the health risks of smoking, by merely showing that the science was uncertain and thereby raising sufficient doubts to hold off regulations (Michaels 2008; Oreskes and Conway 2010). Industry executives knew that they did not have to deny the science about the health effects of smoking or establish that cigarettes were not harmful. Rather, they only had to draw attention to the uncertainties in the science and disagreements about how to interpret data. This was accomplished by producing conflicting research, exaggerating uncertainties, and employing a variety of non-scientists to go on television attacking the science, to make the scientific knowledge appear less "settled". Similar tactics have since been used in climate change, and in a variety of other areas such as the toxicity of certain chemicals (Oreskes and Conway 2010; Michaels 2020).

So far, the strategies of discrediting discussed here are aimed at discrediting certain bodies of knowledge. But discrediting can also target certain knowers or epistemic agents. Knowers can be discredited when their status, ability, credibility, or reliability as epistemic agents is undermined or called into question.

Not all instances of discrediting create ignorance. After all, there are instances when it is *reasonable* to call into question the credibility or reliability of a particular individual. For example, sometimes individuals present themselves as "experts" on a topic where their credentials are either irrelevant or non-existent. Dr. Scott Atlas, who served on President Trump's Coronavirus Taskforce had his expertise questioned when he made several claims about the efficacy of masks, or the purported benefits of allowing the virus to run rampant, in order to achieve "herd immunity". The concern was that Atlas had no expertise related to epidemiology, infectious diseases, or any other field that might be relevant to evaluating the evidence for these claims. His background and training were in radiology. Yet in some cases, individuals are dismissed or discredited, not because they lack relevant expertise, but because they belong to some group or possess some epistemically irrelevant feature (such as their sex, gender identify, or ethnicity). We have already seen that Indigenous fossil knowledge, Indigenous knowledge about ecology, and conservation biology was dismissed by Anglo researchers beginning in the 18th century, in part because they were viewed as "savages" that were not full epistemic agents or knowers. Women, and particularly women of color, were historically denied educational opportunities because they were viewed as not capable of either understanding or fully producing knowledge. Thus, discrediting occurs when the experiences and claims of entire groups are dismissed in virtue of belonging to a group that is viewed as epistemically inferior.

Discrediting on the basis of irrelevant features can also happen in subtler ways. As the MeToo movement showed, many survivors of sexual assault were not taken seriously. Some survivors are not believed at all, their motives are questioned, and in some cases, their own experiences are called into doubt. When Christine Blaise Ford testified that then Supreme Court nominee Brett Kavanaugh had sexually assaulted her when they were both in high school, members of the Senate, who subsequently upheld Kavanaugh's nomination, dismissed the credibility of her testimony. Some of them argued that they believed she had been attacked by *someone*, but doubted the reliability or accuracy of her memories in thinking it had been Kavanaugh. Incest survivors, including Dylan Farrow, were dismissed for having false memories – even when there is strong corroborating evidence of their claims. This sort of discrediting is also thought to involve a certain kind of injustice to such knowers – which will be discussed in greater detail in the next chapter.

Finally, discrediting can be directed, not toward a particular epistemic agent or even against a particular claim, but against institutions or information sources. Claims about "fake news" are often geared toward aiming to generate distrust of certain kinds of sources (e.g., scientists, the government, or "mainstream" media). When trust is undermined, then people may not know what to believe in any particular case (de Melo Martín and Intemann 2018). Ignorance can be created because we are deprived of sources that we believe are trustworthy or reliable.

Discrediting can occur in a variety of ways, but the consequence is that ignorance is created or maintained because evidence or testimony that *should* be taken seriously is unfairly dismissed. Discrediting strategies, however, are complex precisely because they mimic reasonable epistemic practices. That is, in many contexts it is *reasonable* to evaluate whether testimony is credible, or whether the agent might have conflicts of interest, relevant expertise, or some other personal or political motivations that might make their testimony questionable. Distinguishing what counts as "reasonable" or "unreasonable" criticisms of testimony is quite challenging, particularly because those that engage in discrediting can attempt to make such criticisms based on epistemically relevant factors (even if their motives are not epistemically pure). Thus, identifying when, or in which cases, discrediting creates or produces ignorance is difficult and may be an important area of future research.

Disinformation

As alluded to in the discussion of discrediting, ignorance can also be produced and maintained by the dissemination of disinformation or outright false or fabricated claims. While attempts to generate false beliefs are not new, what has changed is the ease and scale at which they can be generated. Social media has made it possible to disseminate false information very quickly to a large number of people all over the globe. Completely fabricated stories, websites, computer generated images or videos can be shared and widely disseminated in seconds and shared or forwarded even once social media companies try to flag the information as false. Disinformation can also be amplified in a context where there is a lack of trust. Social media circles often involve networks or groups that people belong to because they either have trusted relationships or share certain values. Algorithms that govern social media platforms exploit these networks so that our news feeds often target and limit our information consumption to what the algorithms predict will be appealing to us given our search histories or previous history of clicking, reacting (e.g., liking), or sharing. (For a more detailed discussion of how this occurs, see O'Connor and Weatherall 2019). In addition, our own cognitive biases tend to lead all of us to information and sources that support our own values and view of the world. We are all susceptible to things like confirmation bias – we tend to give more weight to stories or information that confirms our already held beliefs, while discounting evidence that counters them. These ingredients (cognitive biases, social media algorithms, and technology that enables fabrication of information and evidence) can lead us into information silos, echo chambers, or what C. Thi Nguyen (2020) refers to as "epistemic bubbles", where access to counter evidence or contrasting views is limited or even actively discouraged.

Selective representation

Not all instances of ignorance are produced or maintained with disinformation or false beliefs. It is also possible to be misled with partial or selective truths. Teenagers might mislead their parents by selectively telling truths (e.g., this party will have adult supervision), while omitting other relevant truths that are necessary for understanding the context or drawing reliable inferences from those truths (e.g., the "chaperone" is my friend's 21-year-old sibling). Selective representation occurs when relevant facts are obscured or hidden when presenting other claims that are true. Pharmaceutical companies have been found to do this in testing drugs. For example, it is true that studies found that breast cancer patients who took a particular drug for five years had a 95 percent survival rate. They might not tell us, however, that it is also true that the survival rate was still 95 percent when patients only took the drug for two years. This would lead patients to draw the (incorrect) conclusion that they should take (and pay!) for a drug for five years, instead of only two. Similarly, chemical companies funded studies purporting to show a reduction in testicular cancer in rats who were exposed to low doses of heavy metals. While it was again true that there was a reduction in particular forms of cancer with some exposure, it gave the misleading perception that low doses of heavy metals are actually beneficial for men. What those studies failed to tell us (and failed to check) is that risks of *other* kinds of cancer still increased at these same low doses. Political ads operate in a similar way. They are often misleading because they can take certain words or votes out of context. They don't show us the whole picture because they are trying to lead us to a particular conclusion.

Again, these strategies are not intended to be an exhaustive list of how ignorance is produced and maintained. Nonetheless, they present a picture of the multiple ways in which ignorance is produced (whether with good or bad intentions).

Conclusions

The study of ignorance as a significant area of feminist scholarship grew from several sources. Feminists, critical race and disability theorists, and decolonial theorists recognized that there was some knowledge (particularly about the experiences of marginalized groups) that did not receive uptake by dominant groups. Thus, they were motivated to theorize and understand why certain experiences were not widely known. Yet, as we shall see in more detail in the next chapter, these scholars also recognized that knowledge is power. The creation and maintenance of ignorance is an assertion of power that can be harnessed to maintain the status quo. The strategies discussed in this chapter can be employed, either intentionally or unconsciously, in ways that make it difficult to recognize, understand, or change institutions, policies, or experiences. It can deprive epistemic agents or groups of resources.

The recognition that ignorance could be the product of social forces raises many further questions. As noted throughout this chapter, ignorance can be problematic or helpful. When is ignorance harmful and what kinds of harms does it bring about? Perhaps most importantly, insofar as ignorance is harmful, what can be done to address it (especially when it may be the result of forces that are unconscious or unintentional)? These questions will be taken up in the next chapter.

Discussion questions

1. Why might epistemologists and philosophers of science be interested in exploring ignorance?

2.   Some sorts of ignorance may, in fact, be desirable. What are some examples? Other forms of ignorance can be harmful. What are some examples of these?
3.   Think of some examples of the types of ignorance that appear in Tuana's taxonomy of ignorance. Which types of ignorance might be of particular interest to feminist epistemologists and why?
4.   What are some examples of how knowledge can be discredited?
5.   Why is the study of ignorance, how it is produced and maintained, of interest to feminist epistemologists and philosophers of science?

Additional suggested readings

Code, L. (2020). *Manufactured uncertainty: Implications for climate change skepticism.* State University of New York Press. An account of epistemic responsibility in response to uncertainty and doubt.

Kourany, J., and M. Carrier. (eds.). (2020). *Science and the production of ignorance: When the quest for knowledge is thwarted.* Cambridge, MA: MIT Press. A collection of essays on the different ways in which ignorance is maintained and produced in science.

Mills, C. (2007). White ignorance. *Race and epistemologies of ignorance,* 247: pp. 26–31.

Proctor, R. N., and L. Schiebinger (eds.). (2008). *Agnotology: The making and unmaking of ignorance.* Stanford, CA: Stanford University Press. A collection of essays addressing multiple aspects of ignorance discussed here.

Sullivan, S., and N. Tuana (eds.). (2007). *Race and epistemologies of ignorance.* Binghamtom, NY: SUNY Press. A collection of essays addressing types and strategies for producing and maintaining ignorance, particularly White ignorance.

References

Alcoff, L. Martín. (2007). Epistemologies of ignorance: Three types. In: S. Sullivan and N. Tuana, (eds.), *Race and epistemologies of ignorance.* Albany, NY: SUNY Press.

Arditti, R. (1977). Have you ever wondered about the male pill? In: C. Dreifus (ed.), *Seizing our bodies: The politics of women's health.* New York: Vintage Books.

Baldwin, J. (1985). No name in the street. In: J. Baldwin, *The Price of the ticket: Collected non-fiction 1948–1985.* New York: St. Martin's Press.

Bottomiller Evich, H. (2016). The revenge of the rural voter. *Politico.* November 13, 2016. https://www.politico.com/story/2016/11/hillary-clinton-rural-voters-trump-231266.

Dalmiya, V., and Alcoff, L. (1993). Are 'old wives' tales' justified? In: L. Alcoff and E. Potter (eds.), *Feminist epistemologies.* New York: Routledge, pp. 217–244.

de Melo-Martín, I., and K. Intemann. (2018). *The fight against doubt: How to bridge the gap between scientists and the public.* Oxford: Oxford University Press.

Farrow, D. (2017). Op-Ed: Why has the #MeToo revolution spared Woody Allen? *Los Angeles Times*, December 7, 2017. https://www.latimes.com/opinion/op-ed/la-oe-fa rrow-woody-allen-me-too-20171207-story.html.

Fernández Pinto, M. (2015). Tensions in agnotology: Normativity in the studies of commercially driven ignorance. *Social Studies of Science*, 45 (2), pp. 294–315.

Fernández Pinto, M. (2020). Open science for private interests? How the logic of open science contributes to the commercialization of research. *Frontiers in Research Metrics and Analytics*, 5:588331.

Frye, M. (1983). *The politics of reality: Essays in feminist theory.* New York: Crossing Press.

Gardner, C. (2017). Blake Lively on Harvey Weinstein Claims: It is devastating to hear. *The Hollywood Reporter*, October 11, 2017. https://www.hollywoodreporter.com/ news/general-news/blake-lively-addresses-harvey-weinstein-allegations-devastatin g-hear-1047599/.

Gilson, E. (2011). Vulnerability, ignorance, and oppression. *Hypatia: A Journal of Feminist Philosophy*, 26 (2), pp. 308–332.

Harding, S. (2008). *Sciences from below: Feminisms, postcolonialities, and modernities.* Durham, NC: Duke University Press.

Harding, S. (2015). *Objectivity and diversity.* Chicago: University of Chicago Press.

Harding, S., and Mendoza, B. (2021). Latin American decolonial feminist philosophy of knowledge production. In: S. Crasnow and K. Intemann (eds.), *The Routledge handbook of feminist philosophy of science*, New York: Routledge, pp. 104–116.

Kourany, J. A. (2015). Science: For better or worse, a source of ignorance as well as knowledge. In: M. Gross and L. McGoey (eds.), *Routledge international handbook of ignorance studies.* New York: Routledge, pp. 155–168.

Le Morvan, P. (2010). Knowledge, ignorance and true belief. *Theoria*, 77 (1): pp. 32–41.

Lugones, M. (2003). *Pilgrimages/Peregrinajes: Theorizing coalition against multiple oppressions.* Lanham, NJ: Rowman & Littlefield Publishers.

Mayor, A. (2008). Suppression of Indigenous fossil knowledge: From Claverack, New York, 1705 to Agate Springs, Nebraska, 2005. In: R. N. Proctor and L. Schiebinger (eds.), *Agnotology: The making and unmaking of ignorance.* Stanford: Stanford University Press, pp. 163–182.

Mayor, A. (2011). *The first fossil hunters: Dinosaurs, mammoths, and myth in Greek and Roman times.* Princeton: Princeton University Press.

Medina, J. (2013). *The epistemology of resistance: Gender and racial oppression, epistemic injustice, and the social imagination.* Oxford: Oxford University Press.

Medina, J. (2017). Epistemic injustice and epistemologies of ignorance. In: P. Taylor, L. Alcoff, and L. Anderson (eds.), *The Routledge companion to philosophy of race.* New York: Routledge, pp. 247–260.

Michaels, D. (2008). *Doubt is their product: How industry's assault on science threatens your health.* Oxford: Oxford University Press.

Michaels, D. (2020). *The triumph of doubt: Dark money and the science of deception.* Oxford: Oxford University Press.

Mills, C. (1997). *The racial contract.* Ithaca, NY: Cornell University Press.

Nguyen, C. T. (2020). Echo chambers and epistemic bubbles. *Episteme*, 17 (2): pp. 141–161.

Nicholas, G. (2018). It's taken thousands of years, but Western science is finally catching up to traditional knowledge. *The Conversation*, February 15. https://the conversation.com/its-taken-thousands-of-years-but-western-science-is-finally-ca tching-up-to-traditional-knowledge-90291.

O'Connor, C., and. J. O. Weatherall. (2019). *The misinformation age*. New Haven: Yale University Press.

Oreskes, N. and E. M. Conway. (2010). *Merchants of doubt: How a handful of scientists obscured the truth on issues from tobacco smoke to global warming*. New York: Bloomsbury Publishing.

Ortega, M. (2006). Being lovingly, knowingly ignorant: White feminism and women of color. *Hypatia: A Journal of Feminist Philosophy*, 21 (3), pp. 56–74.

Oudshoorn, N. (2003). *The male pill: A biography of a technology in the making*. Durham, NC: Duke University Press.

Peels, R. (2011). Ignorance is lack of true belief: A rejoinder to LeMorvan. *Philosophia*, 39 (2): pp. 345–355.

Proctor, R. N. (1995). *Cancer wars: How politics shapes what we know and don't know about cancer*. New York: Basic Books.

Proctor, R. N. (2008). Agnotology: A missing term to describe the cultural production of ignorance (and its study). In: R. N. Proctor and L. Schiebinger. *Agnotology: The making and unmaking of ignorance*. Stanford: Stanford University Press, pp. 1–35.

Proctor, R. N. (2011). *Golden holocaust: Origins of the cigarette catastrophe and the case for abolition*. Berkeley, CA: University of California Press.

Ryzik, M. (2017). Kate Winslet relieves two haunting experiences. *New York Times*, September 6, 2017. https://www.nytimes.com/2017/09/06/movies/kate-winsle t-relives-two-haunting-film-experiences.html.

Schiebinger, L. (2017). *Secret cures of slaves: People, plants, and medicine in the eighteenth-century Atlantic world*. Stanford: Stanford University Press.

Setoodeh, R. (2017). Kate Winslet calls Harvey Weinstein allegations 'disgusting and appalling'. *Variety*, October 9, 2017. https://variety.com/2017/film/news/kate-win slet-harvey-weinstein-allegations-sexual-harassment-scandal-1202584733/.

Smithson, M. (1989). *Ignorance and uncertainty. Emerging paradigms*. New York: Springer.

Smithson, M. (2008). Social theories of ignorance. In: R. N. Proctor and L. Schiebinger (eds.), *Agnotology: The making and unmaking of ignorance*. Stanford: Stanford University Press, pp. 209–229.

Spelman, E. (2007). Managing ignorance. In: S. Sullivan and N. Tuana (eds.), *Race and epistemologies of ignorance*. Albany, NY: SUNY, pp. 119–131.

Sullivan, S. & N. Tuana (eds.). (2007). *Race and epistemologies of ignorance*. Albany: SUNY Press.

Tremain, S. (2017). Knowing disability, differently. In: I. J. Kidd, J. Medina, and G. Pohlhaus, Jr. (eds.), *The Routledge handbook of epistemic injustice*. New York: Routledge, pp. 175–184.

Tuana, N. (2004). Coming to understand: Orgasm and the epistemology of ignorance. *Hypatia: A Journal of Feminist Philosophy*, 19 (1), pp. 194–232.

Tuana, N. (2006). The speculum of ignorance: The women's health movement and epistemologies of ignorance. *Hypatia: A Journal of Feminist Philosophy*, 21 (3): pp. 1–19.

Wuthnow , Robert. (2018). *The left behind: Decline and rage in rural America*. Princeton, NJ: Princeton University Press.

Zeitchik, S. (2016). Woody Allen addresses Ronan Farrow fallout (sort of). *Los Angeles Times*, May 12, 2016. https://www.latimes.com/entertainment/envelope/film festivals/la-et-mn-woody-allen-cafe-society-interview-farrow-20160512-snap-story. html.

6

WHY DOES IT MATTER WHAT WE KNOW?

Overview

The central thesis of this chapter is that what we know (and do not know) and who knows it are not only epistemically important, but also important from ethical perspectives and considerations of social justice. This is a pressing concern in societies such as our own, that rely on science to guide policy. We examine how our epistemic practices can result in epistemic injustices by disproportionately harming knowers on the basis of gender, race, and other social categories and reinforcing social inequalities. Epistemic injustice can take a variety of forms. Among those explored in this chapter are testimonial injustice, participatory injustice, hermeneutical injustice, epistemic trust injustice, and the unjust imposition of epistemic risk.

Introduction

On September 27, 2018, Dr. Christine Blasey Ford testified before the US Senate Judiciary Committee that she was 100 percent certain that she was assaulted by then Supreme Court Nominee Brett Kavanaugh and his friend Mark Judge when they were both in high school. She testified about several details of the assault, even though it had occurred many years previously. Judge Kavanaugh later testified that he was 100 percent certain that he did *not* physically or sexually assault Dr. Ford or anyone. Who should we believe? On what grounds should we judge the credibility of testimony or other evidence? What kinds of methods should be used (if any) to corroborate the evidence that we have? How much evidence do we need to believe someone and do higher stakes (e.g., giving someone a lifetime appointment to

DOI: 10.4324/9781032693781-6

the highest court in the US) influence what degree of confidence is needed? The answers to these questions matter – not only in this case but in many instances where we are trying to decide who or what to believe in relation to science, medicine, law, politics, and policy. They matter because the knowledge we produce or fail to produce provides the basis for how we live our lives, the policies and practices that decisionmakers adopt, and the development of strategies or interventions that can shape our safety, the quality of our lives, and the opportunities that we have.

In previous chapters, we have examined several important questions that have shaped feminist epistemology and philosophy of science: *who* knows? *what* do we know? *how* do we know? and what *don't* we know? In this chapter, we examine in more detail how feminists have identified and distinguished the epistemic and social harms that can result from not being attentive to social and political norms, values, and assumptions in the epistemic practices that we have discussed in previous chapters. To be sure, epistemic and scientific practices also have the power to yield many epistemic and social benefits, and how these might be promoted will be addressed in Chapter 7. Our focus here, however, is on how harms can occur from problematic practices so that we might understand how to avoid or minimize them.

Social and epistemic harms

There are a host of social and ethical harms that can result from our knowledge practices. If we do not know that a certain chemical is toxic, we might not protect communities from exposure to that chemical and there might, for example, be a higher rate of cancer as a result. If computer scientists and engineers program facial recognition software using training sets comprised only of White men, certain technology may not work for other groups. This, in turn, can exacerbate certain social inequalities, such as wrongful detainment or incarceration that occurs when people of color are systematically misidentified by a biased facial recognition program. If certain groups were excluded from participating in science, technology, engineering, and mathematics (STEM) disciplines, solely on the basis of their gender or race, it would constitute an unjust form of discrimination. That is, it would result in certain social groups having their opportunities constrained unjustly. If lawmakers rely on information from those with dubious expertise, it may lead to bad policies that are either ineffective in addressing pressing social problems or that have unforeseen negative consequences. What we know and don't know can harm our health, economy, safety, and security, and even our relationships with others. Moreover, the distribution of these harms can be unfair or unjust, when they disadvantage certain groups or reinforce existing social inequalities. These examples we refer to as *social harms* because they are harmful to our individual and social wellbeing. That is, they thwart or erode our ability to achieve certain social or ethical ends or goods, such as flourishing, equality, or justice.

Yet some harms may be distinctly *epistemic*. Epistemic harms are harmful to our ability to produce and achieve knowledge. Disinformation or propaganda may inflict epistemic harm insofar as they lead us to have false beliefs. Censorship or a lack of transparency may lead to epistemic harms when they prevent us from being able to access certain kinds of information or evidence. A failure to investigate certain diseases might prevent us from knowing the causes or mechanisms of those diseases and making decisions about how best to treat or avoid them. Being prevented from having an education might deprive one of the epistemic resources and skills to be able to be an effective knower or epistemic agent. These are examples of epistemic harms because they impede or erode our ability to know or to achieve knowledge.

The distinction between "social" and "epistemic" harms is not a sharp one, as epistemic harms may *also* produce social harms. For example, practices that keep the public ignorant about the risks of certain chemicals might produce false beliefs (such that a chemical is safe or that one need not avoid exposure) but this, in turn, might also cause them to have higher rates of cancer (a social harm to wellbeing). Similarly, certain social harms might also lead to epistemic harms. For example, societies that systematically constrain the rights of women inflict an obvious social harm on women, but this may also deprive them of the ability to participate in knowledge production or fully develop or employ their epistemic capacities. In addition, it might deprive epistemic communities in those societies of the expertise and knowledge of certain groups. Thus, epistemic and social harms can interact. We distinguish between these kinds of harms, however, because it is useful to be attentive to different domains of harm (and how they interact). This chapter will focus on epistemic harms, as these have been of particular interest to the work of feminist epistemologists and philosophers of science (and largely neglected by traditional epistemologists).

Epistemic injustice

Feminist scholars have largely focused on certain kinds of epistemic harms that arise – specifically harms that exploit power imbalances and support or reinforce systems of inequality or oppression. While there are certainly some epistemic harms that can impact all epistemic agents – for example, believing something false – feminists have shown that oppressed groups have experienced certain kinds of epistemic harms precisely because of their membership in those groups. Feminist philosopher Miranda Fricker (2007) coined the term "epistemic injustice" to refer to a particular set of epistemic harms in which someone is wronged specifically in their capacity as a knower in virtue of their membership in some group, such as gender presentation, sexuality, race, ethnicity, or disability. Broadly speaking, epistemic injustices occur when certain groups are systematically denied epistemic resources or are treated as non-epistemic agents (or less than fully human as knowers). Epistemic injustices are a specific subset of epistemic harms that disproportionately impact certain groups who have also been historically marginalized.

While the term "epistemic injustice" is relatively recent, it evolved from a long history of work by feminist, critical race, critical disability, and decolonial scholars. As Vivian May (2014) points out, the idea that individuals or groups could be epistemically wronged or epistemically oppressed can be attributed to Sojourner Truth, who in 1867 highlighted the denial of Black women as knowers (May 2014: 98) and to Julia Anna Cooper, who asserted in 1892 Black women have been "one mute and voiceless note" *not* because they do not have knowledge to contribute, but because they have been denied self-representation in public discourse and willfully misunderstood (Cooper 1892/ 1988; May 2014: 97). Kristie Dotson (2011) and Gaile Pohlhaus, Jr. (2017) also note that a similar idea can be found in Gayatri Chakrovorty Spivak's decolonialist work (1988), which introduced the notion of "epistemic violence" as occurring when subaltern or marginalized persons are prevented from speaking about their own experiences or interests by those who claim to know what those interests are.

The philosophical literature on epistemic injustice has grown quite rapidly in recent years. Here we highlight some of the ways that epistemic injustices can occur, though this list is not intended to be exhaustive and, as some have cautioned, epistemic injustices may not capture all of the serious epistemic harms that might constitute broader systems of "epistemic oppression" (Dotson 2012; Medina 2017).

Types and sources of epistemic injustice

Testimonial injustice

As we saw in Chapter 2, one of the central arguments that feminist scholars have made is that we must rely on others to know. Testimony is one form of evidence that (as we saw in the opening example of Christine Blasey Ford) can play an important role in drawing conclusions and justifying beliefs. It plays an obvious role in legal cases because jurors must rely on the testimony of witnesses, defendants, victims, and experts to determine who is credible and what conclusions to draw about guilt or culpability. Outside of legal contexts, we rely on testimony to form our beliefs about a variety of experiences – precisely because few individuals can experience or have first-hand knowledge of *everything* by themselves. We rely on the testimony of experts to tell us when a drug is safe and effective, whether our cars or homes will need repairs, or whether our water is contaminated. We rely on the testimony of others who write reviews on the internet to tell us whether a particular product or service is one we ought to use. We sometimes rely on the testimony of others to decide how to vote, or what policies to support. Even scientists – who may be experts in some areas – must rely on the testimony of other scientists or technicians to build on previous research that appears to be promising, or to trust that their equipment is functioning properly. In other words, we grant others epistemic or cognitive authority for specific domains of knowledge, and, in some cases, we must trust the testimony of others to know. This makes our decisions and practices about to whom to grant such credibility or authority important. There is a risk that such decisions can be arbitrarily influenced by power, bias (implicit or explicit), and interests. Having unreliable or arbitrary standards about whose testimony to trust can result in less warranted beliefs and epistemic injustices to knowers or groups of knowers.

According to Miranda Fricker, among the central kinds of epistemic injustices that can occur are testimonial injustices. Testimonial injustice happens when someone is ignored or not believed *solely* due to prejudices against certain features, such as sex, sexuality, gender presentation, race, socioeconomic status, or disability (Fricker 2007). Fricker (2012) uses the example of Duwayne Brooks who, in 1993, was with his friend Stephen Lawrence when he was murdered in London. Brooks and Lawrence (who were Black) had been waiting at a bus stop and were approached by a small group of White teenagers who fatally stabbed Lawrence in what was eventually found to be a racially motivated attack. When police arrived at the scene, they failed to recognize Brooks as a witness to a crime – but rather assumed he was a suspect. The senior officer on the scene did not ask him what happened and the officers did not believe him when he tried to explain what had happened. They did not ask for descriptions of the White teenagers he said had been responsible or ask him to help in identifying the suspects in a search of the area, although he knew where they had been seen. An inspector who had been called upon to investigate the police handling of this case concluded that it was the result of institutional racism within the police force and that they did not find Mr. Brooks credible – or treat him as a witness – because of his race. Fricker argues that testimonial injustice occurs when prejudice distorts a hearer's perception of a speaker such that it deflates the credibility given to the speaker.

Testimonial injustice can occur in many contexts, including in science. In the early 1950s, when astronomist Vera Rubin first hypothesized the existence of dark matter to explain observations that were at the time inconsistent with prevailing theories in astronomy, she was told by members of the audience at her American Astronomical Society talk that perhaps she should go home to her children and her paper was never published (Yount, 2006: 126). While there might have been many factors that contributed to the dismissal of her work, it is plausible to think that some did not take her work seriously because she was a woman. Stereotypes that women were "irrational" or "not cut out for science" and norms that she should be "at home" rather than in a lab made it difficult for other male scientists to think she was credible or take her testimony seriously. It took many years, but Rubin's work was finally vindicated. This denial of epistemic authority on the basis of gender, however, potentially hindered or slowed progress in the field of astronomy.

While Fricker's account of testimonial injustice focuses on "the injustice that a speaker suffers in receiving deflated credibility from the hearer owing to identity prejudice on the hearer's part" (Fricker 2007: 4), testimonial injustice may operate in more complex ways. Emmalon Davis argues that in addition to identity-based testimonial injustices that occur in the cases described above, speakers can also suffer credibility deficits in virtue of the *content* or kinds of information that a speaker attempts to convey (Davis 2021). As Davis explains, the content of certain ideas, claims, or theories can become identity-coded "in ways that provoke unwarranted epistemic assessments of both the contributor and the contribution" (Davis 2021: 218). Consider some of the initial reactions to the very idea of feminist epistemology and feminist philosophy of science described in Chapter 1. Imagine a White cisgendered male philosophy professor who tries to explain some feminist critiques of science or philosophy of science to his colleagues, to encourage them to include some of this literature in their classes. Further imagine that he is interrupted by another colleague who says: "Those are just a bunch of angry women who don't like science because they don't like the truths science produces". Even though the credibility of the speaker is not called into question in virtue of *speaker's* identity, his credibility is challenged because the content that he is offering is content that is closely associated with the identity of a larger group against whom there is prejudice on the hearer's part. As in the case of identity-based testimonial injustice, content-based testimonial injustice can lead to the dismissal of the speaker's contribution and credibility. It is distinct, however, because it relies on prejudice against the content that is linked to a group whose credibility is also being undermined (Davis 2021).

In addition, Davis (2016) argues that testimonial injustices do not always involve an unjustly compromised assessment of the speaker's credibility. Such injustices can also result from positive stereotyping that gives rise to prejudicially *inflated* credibility assessments. Stereotypes of marginalized groups can be "positive" in the sense that the attribute is seen as positive or beneficial by a dominant group. For example, that women are nurturing, or that Asian Americans are good at math, or that African Americans excel at athletics are thought to be "positive" or "benevolent" stereotypes. Yet these stereotypes can also cause a variety of harms to members of those groups.

For instance, they may be discouraged from pursuing activities that deviate from those stereotypes or receive harsher judgments if they fail to conform to them. Davis argues that such stereotypes can also influence assessments of speaker credibility, leading to epistemic injustices. For instance, positive stereotypes can cause "typecasting" where members of a marginalized group are taken to be credible, or even expected to be experts, in a very narrow range of subjects that reflect the stereotype (Davis 2016). In this way, even positive stereotypes and related credibility excesses can shape and restrict a marginalized group's role in an epistemic community. It can set them up for failure within epistemic communities because they may not in fact be experts in the domains where it is expected yet also not allowed to be experts in other areas.

Testimonial smothering and anticipatory epistemic injustice

In many of the cases of testimonial injustice described above, testimony is dismissed by the hearer. Kristie Dotson (2011) has argued that there is another kind of epistemic violence that occurs when knowers remain silent or "smother" their own testimony for fear of not being heard, believed, or understood. Testimonial smothering, or the truncating or suppression of one's own testimony can occur when one's audience is unlikely to be willing or able to be receptive to or engage in uptake of the testimony (Dotson 2011: 244). As Dotson points out, communicating requires an audience willing to listen and engage in linguistic reciprocity. But there are cases where there may be a failure of testimonial reciprocity because of "an audience failing to meet speaker dependencies in a linguistic exchange" (Dotson 2011: 239). Veronica Ivy, for example, asks us to imagine the futility of trying to explain the concept of White privilege – or explaining examples of White privilege – to an audience of White supremacists (Ivy 2019: 290). In this case, the speaker may experience coerced silence – or a selective self-censorship due to the perception that an audience lacks the ability to genuinely understand or comprehend such testimony (Dotson 2011: 244). Dotson argues that cases of testimonial smothering tend to happen when there is a significant risk that the testimony will be unintelligible and that this lack of comprehension is also likely to result in the formation of "false beliefs that can cause social, political, and/or material harm" (Dotson 2011: 244). For example, Black women may smother testimony about sexual or domestic violence because it risks reinforcing racist stereotypes about Black men. Indeed, testimonial smothering occurs in many cases of sexual assault because survivors fear that they will not be believed and will likely suffer a variety of personal, social, or professional consequences as a result.

Testimonial smothering can be thought of as a kind of testimonial injustice, in that it involves a silencing of testimony (albeit on the part of the speaker). What distinguishes cases of testimonial smothering (from Fricker's more general account of testimonial injustice) is that they are not the denial of epistemic credibility because of prejudices, but rather self-censorship that occurs because of fear that an audience lacks the ability to understand and the testimony would thus risk social or institutional penalties.

Ji-Young Lee argues that there is a further kind of self-silencing that can occur even when an audience does not in fact lack the sort of testimonial competence needed to comprehend a speaker (Lee 2021: 568–569). Lee argues that cases of *anticipatory epistemic injustice* can occur because a speaker fears their testimony will not receive uptake. Lee considers the example of LGBTQIA+ individuals who may silence themselves from "coming out" or sharing their identity, not because they are afraid that they may not be believed or understood, but because they feel ashamed and worry they will be stigmatized. Lee's account of anticipatory epistemic injustice captures harms that occur not only because of external pressures on marginalized epistemic agents, but also internal forces that may make it more difficult for them to share their experiences. They are silences that occur in anticipation of other epistemic injustices that may occur. According to Lee,

> agents are wronged by this phenomenon both because they are unjustly placed in this position of epistemic precarity, which make them prone to all the affects and uncertainties that manifest as anticipatory behavior, and because they unduly partake in relinquishing the more epistemically authoritative testimonial routes typically available to others.
>
> *(Lee 2021: 570)*

Participatory injustice

Instances of testimonial injustice happen when testimony is ignored or silenced because those who *deserve* to be recognized as having cognitive authority or epistemic credibility are ignored or silenced. Within the practice of science, these issues become more complicated because not every epistemic agent *has* the same degree of cognitive authority or epistemic credibility in a particular research context. Scientific and technical knowledge can be highly specialized and require various kinds of training and expertise, and we simply cannot all be experts in everything. Yet, while not all epistemic agents have expertise in everything, another sort of injustice can occur when certain groups are systemically limited or prevented from achieving some area of expertise in virtue of their membership in a particular social group (race, class, gender, etc.). Heidi Grasswick has referred to these as *participatory injustices* (Grasswick 2017).

Despite significant progress over the years, women, people of color, and persons with disabilities continue to be underrepresented in STEM degrees and in the STEM workforce compared to their representation in the population and the workforce (NSF 2023: 3). Although women and men are represented similarly in the total workforce, women (from all intersecting categories) accounted for 35 percent of STEM workers in 2021 (NSF 2023: 3). Those who had at least one disability were approximately 9 percent of the population the in US, but about 3 percent of the STEM workforce. Collectively, men and women identifying as Black, Hispanic or Latino, American Indian/Alaska Native made up 31 percent of the total US population in 2021 and 24 percent of STEM workers (NSF 2023: 5). Moreover, underrepresented groups within STEM tended to participate in STEM positions that required less education and earned lower median incomes than White male STEM workers (NSF 2023: 3–4). Within STEM, underrepresentation is also particularly persistent in certain fields. For example, there are lower numbers of women in physics, computer science, electrical engineering, and mechanical engineering, even though representation in the life sciences is approaching 50 percent (NSF 2023: 46–7). The numbers are similar in the field of philosophy where about 20 percent of those who identify as philosophers of science also identify as women (Brister and Hicks 2021). Feminist scholars have shown that gender and racial norms and stereotypes have influenced both individual and institutional practices in ways that have created obstacles for women and men of color to gain expertise and authority in STEM fields.

Unjust denial of epistemic authority and expertise can occur through formal or informal barriers to participation in STEM or STEM education. Social conditions can also prevent people from acquiring the means viewed as necessary for acquiring expertise or epistemic authority, such as education. For example, access to resources that facilitate interest and training in STEM disciplines is not equitably distributed among populations (De Welde and Laursen 2011). In the US, one-third of K-12 schools are rural and one in five students attends a rural school, yet rural communities receive less funding for education that their urban counterparts, tend to have less access to technologies, such as high-speed internet, and have a more difficult time recruiting and retaining qualified STEM teachers (Avery 2013). In addition, the content and culture of STEM disciplines can be expressed in ways that are alienating for certain groups. For example, many rural, poor, female, and Indigenous STEM students report a disconnect between academic STEM knowledge and their local knowledge practices, which they see as more relevant to the "real world" and things they care about (Avery and Kassam 2011; Cech et al. 2017). As a result, some underrepresented STEM students drop out because they do not find the content of the disciplines to be inclusive.

Participatory injustice can also occur even among those who are able to become experts. Credentialed scientists who are a part of epistemic communities and research teams can still have their participation and contributions ignored, belittled, or challenged. When members of underrepresented groups are taken less seriously or given less uptake in their intellectual interactions due to implicit biases or prejudice, they suffer participatory epistemic injustices (Grasswick 2017: 317). Implicit gender and race biases have been documented in mentoring, hiring, and tenure decisions, citation practices, and grant evaluation (Moss-Racusin et al. 2012; King et al. 2017; Dworkin et al. 2020; Casad et al. 2021; Eaton et al. 2020). Grasswick argues that "participatory injustice can happen when, due to such implicit biases, members of certain groups are not invited or encouraged to submit work to important conferences or publication venues to the same extent as others, or when their names are simply not thought of when members of the profession are soliciting peer reviews (Grasswick 2017: 317). Thus, participatory injustice occurs when marginalized groups have the significance of their epistemic contributions undervalued, dismissed, or altogether ignored, on the basis of factors that are epistemically irrelevant. It is concern for this sort of injustice that, presumably, led Helen Longino (1990) to call for "equality of intellectual authority", as a condition of objectivity as discussed in Chapter 4.

Barriers to broad participation point to institutional cultures and practices that make it difficult for certain groups (especially women of color, Indigenous women, rural and poor women) to become experts in STEM or be recognized as having epistemic credibility. This constitutes an epistemic harm to such potential knowers, but also can harm epistemic communities more generally insofar as they may lack the full talent, experiences, and insights that result from having greater epistemic diversity.

Hermeneutical injustice

Miranda Fricker (2007) also identifies hermeneutical injustice as a type of epistemic injustice, which occurs when a gap in collective interpretive resources (such as language or conceptual frameworks) unfairly deprives certain groups of the ability to interpret or make sense of their own social experiences (Fricker 2007: 1). Consider the binary two-sex model that guided several areas of science discussed in Chapters 3 and 4. For decades, both in scientific research and in much of our social life, our conceptual frameworks explicitly assumed the existence of two (mutually exclusive) sexes (male/female), that tracked onto two genders (masculine/feminine), and also assumed two sexualities ("homosexual" and "heterosexual"). These conceptual frameworks give certain experiences meaning in virtue of how beings and identities can be categorized or described in relation to them. Yet notice that these particular ways of giving meaning make understanding other kinds of experiences impossible. Trans and non-binary persons, for example, lacked the conceptual resources to articulate, identify, or understand their own experiences under the two-sex, two-gender, two-sexualities framework (see the discussion of Sari van Anders Sexual Configuration Theory (SCT) in Chapter 3 as a response to this conceptual insufficiency).

Gaile Pohlhaus, Jr. (2012) emphasizes that hermeneutical injustice is not just about depriving certain groups the interpretive resources for them to understand their own experiences, but also involves a lack of willingness by dominant groups to allow their interpretations of the world to be revised and informed by those who are marginally situated (Pohlhaus, Jr. 2012: 722). Negative reactions to attempts to use more inclusive language and concepts about sex, gender, gender identity, and sexual orientation are one example of how those who are trans, non-binary, or queer, for example, are not only denied epistemic resources for interpreting their own experiences, but also face willful resistance to their criticism of dominant conceptual frameworks (Hall 2017). Indeed, the establishment and maintenance of dominant frameworks can also involve an exercise of power. Some feminist and queer theorists have built on the work of French philosopher Michel Foucault, approaching our concepts of sexuality and sexual identity as the products of the operation of power, rather than innate truths about the world (Foucault 1990; Hall 2017). Foucault used a genealogical approach to sexuality, which examines the historical conditions that gave rise to certain concepts or social beliefs in order to identify the conditions (often involving the exercise of power) that make them possible. In Fricker's terminology, these conditions can result in hermeneutical injustice.

José Medina (2017) argues that some forms of hermeneutical injustice can result in what he calls "hermeneutical death", which is when hermeneutical injustice is so deep and pervasive that it radically constrains one's interpretive capacities to such a degree that it results in the total curtailment of one's voice, or one's ability to participate in meaning-making or meaning-sharing practices. For example, between 1869–1960s, colonial policies in the US and Canada removed Indigenous children from their homes and sent them to boarding schools to be "assimilated". The goal of such policies was cultural genocide or the systematic destruction of Native cultures and communities (Woolford 2015). As a part of those policies, the speaking of Native languages was forbidden, and children were forced to learn and speak English. In these cases, Indigenous children were quite literally stripped of all their ability to participate in meaning-making and meaning-sharing practices about their experiences or culture. Similarly, Gloria Anzaldúa describes Anglo White attacks on Chicano/a language and culture as a form of "linguistic terrorism" and asserts that such "repeated attacks on our native tongue diminished our sense of self" (Anzaldúa 1987: 80). It is this sort of phenomenon that has contributed to calls for the development of emancipatory conceptual frameworks for, as Black feminist Audre Lorde (1984) has argued, "the master's tools will never dismantle the master's house".

Chapter 3 offered some examples of feminist researchers working to alter concepts, so that there is a better fit between the language available and the experiences and phenomena that they seek to give an account of. In doing so, they are pushing back against the constraints of their disciplinary conceptual framework by finding alternatives that allow them to describe the world as they experience it. We might think of them as seeking to redress hermeneutical injustice.

Epistemic appropriation

Emmalon Davis (2018) has identified another kind of epistemic injustice that is distinct from those already mentioned. Epistemic appropriation occurs when epistemic resources developed by marginalized groups receive uptake by dominant groups, but that they are contributions from the marginalized is not acknowledged and they are subsequently used in ways that benefit the interests of the dominant group. Consider, for example, the ways in which Indigenous fossil knowledge was neglected and suppressed (briefly discussed Chapter 5). As Adrienne Mayor (2005) details, remains of Pleistocene mastodons and mammoths as well as other large mammals were encountered by Native observers in northeastern America, long before contact with European colonizers. These fossils were featured in the centuries-long oral traditions of the Iroquois, Delaware, Shawnee, Wyandot, and many other Native peoples as evidence of creatures that had lived and become extinct before the existence of present-day humans (Mayor 2008: 165). At the time, no scientific theory existed in Europe that would explain these bones. During the 18th century, several settler colonist intellectuals became interested in the fossils and the Indigenous knowledge about them. For example, in 1712, clergyman Cotton Mather took a keen interest in giant bones that had been found in Claverack, NY, and wrote about them extensively to the Royal Society of London. While Mather referred to Native peoples as "savages" and "devil worshippers", he appeared to recognize their oral traditions as important evidence and even learned Algonquin in order to access their knowledge. Yet in his writings, while drawing on Indigenous knowledge about the fossils, he took pains to ridicule their language and oral traditions, detaching his theorizing from them (Mayor 2008: 167). Moreover, subsequent Anglo intellectuals that investigated and developed the field of paleontology gradually (and sometimes intentionally) erased the contribution of Native peoples to this body of knowledge. This is an example of what Davis (2018) refers to as "detachment", where the epistemic contributions of marginalized people are detached from them and attributed to others. Moreover, the knowledge was subsequently published by European academics who received the epistemic credit, scientific prestige, grants, and other accolades. Not only did the knowledge fail to benefit Native peoples, but Anglo and European scholars also regularly used their publications to deride these peoples as "primitive", "unscientific", and bound to "myths" and "legends". In other words, not only were their contributions erased, but those very contributions were given credit to others who simultaneously undermined the status of the originators as epistemic agents. Both detachment and the use of marginalized epistemic contributions in ways that fail to benefit those groups constitutes epistemic appropriation in Davis's view.[1]

Epistemic appropriation is distinct from other forms of epistemic injustice discussed above. Unlike testimonial injustice, it does not involve a complete dismissal or suppression of the testimony of marginalized groups (Davis 2018: 721). Indeed, there is uptake of the epistemic contributions of the marginalized by dominant groups. However, those contributions become detached and used in ways that serve the interests of those in power. It is also distinct from cases of testimonial smothering, because it does not (always) involve cases where marginalized groups truncate their own testimony. Finally, it is distinct from cases of hermeneutical injustice because it does not involve a conceptual deficit in shared epistemic resources (Davis 2018: 719). The fossil case is one where the conceptual resources were able to travel across epistemic communities, but the injustice occurs in the ways they are detached from their original contributors and used to reinforce inequalities.

Epistemic trust injustice

Heidi Grasswick (2017) has identified another kind of epistemic injustice that can occur in contexts where laypersons must trust experts (scientists, engineers, or medical professionals). As mentioned earlier, all of us are non-experts with respect to some (if not all) scientific fields. In order to both have, and benefit from, scientific knowledge, laypersons must trust relevant experts. Trusting anyone, including experts, is risky because we make ourselves vulnerable to being let down by those who we trust and, as a result, believing incorrectly and potentially basing our decisions and actions on bad information. Thus, we also must make decisions about *who* to trust or, as Grasswick puts it, where to "place our trust responsibly" (Grasswick 2017: 319). Epistemic trust injustices occur, according to Grasswick, when "due to forces of oppression, the conditions required to ground one's trust in experts cannot be met for members of particular subordinated groups" (Grasswick 2017: 319). This happens when scientific communities and practices have themselves contributed to the oppression of certain groups. Consider, for example, the well-known Tuskegee study of untreated syphilis in African American males (who did not give informed consent to participate) (see, for example, Brandt 1978). Researchers from US Public Health Services, working with the Tuskegee Institute, told the men that they were being given free medical examinations for "bad blood", when in fact they were being examined to study the long-term effects of untreated syphilis. By 1943, penicillin was widely available as an effective treatment for syphilis, but the participants in the study were not offered this treatment and the study continued until it was exposed by reporters in 1972. This, of course, is just one instance where African American communities were studied, without informed consent, but there are many others. The history of experimenting on African Americans without their consent, misleading, or lying to them without regard for their wellbeing has undermined trust in science and healthcare for many African Americans today. Trust requires that laypersons believe that experts genuinely care about one's wellbeing (Baier 1986; de Melo-Martín and Intemann 2018; Goldenberg 2021). Thus, incidents like Tuskegee give African Americans (for example) good reason to withhold trust from experts and institutions who have historically not been reliable in this way.

The undermining of trust also creates an epistemic injustice, because it can disproportionately deny marginalized groups access to epistemic resources. Laypersons must trust experts, but when conditions have been created so that marginalized laypersons cannot *rationally* trust certain experts (because of their history of abusing power, discriminating against certain groups, or failing to care about the wellbeing of marginalized persons), they are deprived of expertise that is needed to be able to have or make use of existing knowledge. Consequently, marginalized laypersons lack access to trustworthy expertise – an epistemic resource often necessary for knowledge – in a way that dominant groups do not. For example, although the COVID-19 vaccine was shown to be safe and effective and was endorsed by the Centers for Disease Control and Prevention (CDC), some African American communities had lower vaccination rates and expressed reluctance to trust the CDC or to use a vaccine that did not have a long history of testing (on others) (Khubchandani and Macias 2021). Because they had good reason to be skeptical of the credibility of medical scientists (given the history described above), they were deprived of expert knowledge as well as health benefits that might come with vaccination. Indeed, one of the strategies that aimed to address this was to involve African American scientists, doctors, and public health officials in the creation of, and messaging about, the COVID-19 vaccine, to ensure that community would have credible experts.

Unjust imposition of epistemic risks

Another kind of epistemic injustice that might occur in science involves systematic unjust imposition of epistemic risks on particular groups. Justin Biddle and Quill Kukla argue that there are a variety of epistemic risks that are involved in scientific decision-making (Biddle and Kukla 2017). Epistemic risks are, broadly speaking, any risk of epistemic errors that arises in knowledge producing practices. We have already seen an example of one kind of epistemic risk in Chapter 4 – namely *inductive risk* that occurs in deciding how much evidence is needed to stop data collection or in deciding to draw a conclusion based on current evidence. In cases of inductive risk, there is always a risk of accepting a false hypothesis or rejecting a true hypothesis. Each of these possibilities can carry both epistemic and social consequences. For example, if we reject the hypothesis that anthropogenic climate change is likely to lead to catastrophic impacts and it turns out to be true, we run the risk of failing to enact policies that might mitigate those consequences. If we accept the hypothesis that anthropogenic climate change is likely to lead to catastrophic impacts and it turns out to be false, we run the risk of overregulating certain industries and behaviors in ways that may have significant economic impacts. Similarly, if we accept that women with high testosterone are unnaturally better athletes than those whose testosterone measures within the average (for women) range and this turns out to be false and we disqualify them from competing on those grounds, then we unfairly prevent them from competing. But if we deny the hypothesis and it is true then we prejudice women whose testosterone falls within the average range.[2] Calculating inductive risks requires evaluating which set of consequences is worse, and which risks are more acceptable.

Biddle and Kukla argue that there are other kinds of epistemic risks that might be involved in epistemic practices, including scientific practices. There are, for instance, what they call ethical epistemic risks at stake in decisions about what areas of research to pursue (Biddle and Kukla 2017: 220). For example, research on race or sex differences might yield valuable information, but such research programs also run the risk of reinforcing deeply held stereotypes or being used to justify sexist and racist policies (Leuschner and Pinto 2021, 2022). Biddle and Kukla also identify "phronetic risks", which are epistemic risks that occur in relation to decisions that are preconditions for empirical reasoning (Biddle and Kukla 2017: 221). As we saw in Chapter 3, conducting empirical research requires adopting particular conceptual schemes and operationalizing concepts, which might be done in different ways that may have different epistemic and social consequences. Similarly in Chapter 4, we saw that there are a variety of methodological choices that must be made, including decisions about how to collect data, what data is relevant, and whether any data should be excluded. These decisions also potentially carry risks because of the consequences they can have.

Because a variety of epistemic choices and decisions involve risks, it raises the question: what is a *fair or just* distribution of risks? To put this another way, *who* should carry the burden of such epistemic risks? Going back to the example of anthropogenic climate change, consider again whether it is *worse* to risk catastrophic climate impacts or overregulation of certain kinds of industries? It is difficult to evaluate this question unless we also consider *who* is at risk in each case. Climate change impacts are more likely to pose devastating impacts to particular groups who have been historically marginalized including poor communities, communities of color, and Indigenous communities. On the other hand, overregulation may impose risks to different populations (depending on the details of the regulation) including corporations, industries, and developed nations. The concern here is that certain epistemic risks of error may disproportionately impact marginalized groups in ways that perhaps could also constitute a kind of epistemic injustice. Similar considerations apply in the case of the athletes. It is typically intersex individuals or trans folk who have been targeted by policies that prohibit high testosterone individuals from competing. These are members of marginalized and oppressed groups already and this seems a factor that should be weighed in judging risk. To our knowledge, scholars working in this area have not identified this as an epistemic injustice, per se, however, if epistemic risks are made to fall largely on marginalized groups, this may constitute a further kind of epistemic injustice.

How can we address or prevent epistemic injustices?

In the last section, we identified several different ways that epistemic injustices can arise, including ways that testimony might be unfairly silenced or dismissed and ways that our collective interpretive resources (concepts, language, frameworks) might be impoverished and result in disproportionate epistemic harms to particular groups. Many of the epistemic harms and injustices that have been identified in this chapter are not always intentional. Indeed, many of them are caused by neglect, lack of awareness of hidden shared assumptions or values, limitations of experiences, or cognitive limitations that we have as humans. This raises several important questions. First, if they are not intentional, are individuals or epistemic communities *responsible* for such injustices? After all, lack of awareness, the limitations of experiences, and cognitive limitations are not things that we can fully control. Second, if they are unintentional, what can be done to address them?

In Chapter 4, we saw that while it was difficult for individuals to recognize their own idiosyncratic biases, this was potentially easier to do at the level of communities, particularly communities that are structured in certain ways to achieve "procedural objectivity". While individual scientists might not realize when their reasoning is flawed, that can be caught and corrected if there is a diverse group of scientists who can engage in transformative criticism of that reasoning.

It might be the case that epistemic injustices could be addressed in similar sorts of ways. If we are aware of how epistemic injustices can occur, we can ask: how can we change or shape our epistemic practices to avoid such harms? We might try to identify, for example, practices and mechanisms for trying to prevent or correct epistemic injustices. What exactly those practices might look like may depend on the different ways that epistemic injustices occur. For example, addressing participatory injustices will require examining and identifying the barriers to STEM education (cultural, economic, social, and institutional) and adopting strategies for dismantling them. Preventing certain kinds of testimonial injustices may require better understanding and mechanisms for addressing implicit biases. Preventing hermeneutical injustices may require iterative, critical reflection on the kinds of frameworks, concepts, and language we use and explicitly considering what it captures, what it does not, and whether it best serves diverse interests. It may require encouraging and pursuing pluralistic approaches to science, where different sorts of theories and frameworks and ways of thinking about the world are embraced. It might also require encouraging various kinds of diversity in epistemic communities or ensuring that diverse interests, needs, experiences, and ideas are represented. Feminist approaches motivate and aid in developing these sorts of practices.

As we will discuss in the next chapter, there is still quite a bit of work to be done in understanding the different ways in which epistemic injustices can occur, and what can be done about them. This is a promising area for future research and scholarship in feminist epistemology.

Conclusions

In this chapter, we have argued that who knows, what we know, how we know, and what we don't know matters, because our epistemic practices can lead to epistemic harms. In some cases, those harms can also constitute epistemic injustices, insofar as they disproportionately impact certain groups arbitrarily and in ways that may exacerbate existing social inequalities. Just being aware of how epistemic injustices can occur can help us ask: how can we change or shape our epistemic practices to avoid such harms and know "better"? That is, how can we minimize our epistemic limitations and produce knowledge that has a greater chance of improving our lives and addressing inequality and injustice? What epistemic or scientific practices might lead to greater social and epistemic benefits while minimizing harms? As we have seen in Chapter 5, ignorance of various sorts may hinder our ability to do this, as do the sorts of epistemic injustices we have discussed in this chapter. Feminist epistemology has focused attention on these previously unexplored factors that can harm knowledge production. In this way, feminist critique provides an opening to explore how practices of knowledge production might be better able to address the needs and interests of diverse epistemic communities. This will be the focus of the final chapter.

Discussion questions

1. What is epistemic injustice?
2. Why do we need to rely on the testimony of others? Why do scientists need to rely on the testimony of others?

3. What is one example of epistemic injustice discussed that can occur unintentionally?
4. How do questions of inductive risk interact with issues of epistemic injustice?

Additional suggested readings

Kidd, I. J., J. Medina, and G. Pohlhaus, Jr. (2019). *The Routledge handbook on epistemic injustice.* New York: Routledge.

Medina, J. (2012). *The epistemology of resistance: Gender and racial oppression, epistemic injustice, and resistant imaginations.* Oxford: Oxford University Press.

Mills, C. (2002). *The racial contract.* Ithaca: Cornell University Press.

Tremain, S. (2017). *Foucault and feminist philosophy of disability.* Ann Arbor: Michigan University Press.

Notes

1 See Davis 2018 for a detailed discussion of additional examples of epistemic appropriation that can be unintentional (or even driven by apparently "good" intentions), as she illustrates with the case of John Stuart Mill's epistemic appropriation of Harriet Taylor Mill's work and Harriet Beecher Stowe's epistemic appropriation of Sojourner Truth's contributions.
2 The complicated relationship between levels of testosterone and athletic performance is discussed in motre detail in Chapter 7 of Jordan-Young and Karkazis 2019.

References

Anzaldúa, G. (1987). *Borderlands/La frontera: The new mestiza.* New York: Aunt Lutte Books.

Avery, L. (2013). Rural science education: Valuing local knowledge. *Theory into Practice*, 52 (1), pp. 28–35.

Avery, L. M., and K. A. Kassam. (2011). Phronesis: Children's local rural knowledge of science and engineering. *Journal of Research in Rural Education*, 26, pp. 1–18.

Baier, A. (1986). Trust and antitrust. *Ethics*, 96 (2), pp. 231–260.

Biddle, J. B., and R. Kukla, (2017). The geography of epistemic risk. In: K. C. Elliott and T. Richards (eds.), *Exploring inductive risk: Case studies of values in science.* New York: Oxford University Press, pp. 215–237.

Brandt, A. M. (1978). Racism and research: The case of the Tuskegee Syphilis Study. *Hastings Center Report*, 8 (6), pp. 21–29.

Brister, E., and D. J. Hicks. (2021). Contributions of women to philosophy of science: A bibliometric analysis. In: S. Crasnow and K. Intemann (eds.), *The Routledge handbook of feminist philosophy of science.* New York: Routledge, pp. 65–75.

Casad, B. J., J. E. Franks, C. E. Garasky, M. M. Kittleman, A. C. Roesler, D. Y. Hall, and Z. W. Petzel. (2021). Gender inequality in academia: Problems and solutions for women faculty in STEM. *Journal of Neuroscience Research*, 99 (1), pp. 13–23.

Cech, E. A., A. Metz, J. L. Smith, and K. DeVries. (2017). Epistemological dominance and social inequality: Experiences of Native American science, engineering, and health students. *Science, Technology, & Human Values*, 42 (5), pp. 743–774.

Cooper, A. J. (1892/1988). *A voice from the south by a black woman of the south.* New York: Oxford.

Davis, E. (2016). Typecasts, tokens, and spokespersons: A case for credibility excess as testimonial injustice. *Hypatia: A Journal of Feminist Philosophy*, 31 (3), pp. 485–501.

Davis, E. (2018). On epistemic appropriation. *Ethics*, 128 (4), pp. 702–727.

Davis, E. (2021). A tale of two injustices: Epistemic injustice in philosophy. In: J. Lackey (ed.), *Applied Epistemology*. Oxford: Oxford University Press, pp. 215–250.

De Melo-Martín, I. and K. Intemann. (2018). *The fight against doubt: How to bridge the gap between scientists and the public.* Oxford: Oxford University Press.

De Welde, K., and S. Laursen. (2011). The glass obstacle course: Informal and formal barriers for women Ph.D. students in STEM fields, *International Journal of Gender, Science and Technology*, 3 (3), pp. 571–595.

Dotson, K. (2011). Tracking epistemic violence, tracking practices of silencing. *Hypatia: A Journal of Feminist Philosophy*, 26 (2), pp. 236–257.

Dotson, K. (2012). A cautionary tale: On limiting epistemic oppression. *Frontiers: A Journal of Women Studies*, 33 (1), pp. 24–47.

Dworkin, J. D., K. A. Linn, E. G. Teich, P. Zurn, R. T. Shinohara, and D. S. Bassett. (2020). The extent and drivers of gender imbalance in neuroscience reference lists. *Nature Neuroscience*, 23 (8), pp. 918–926.

Eaton, A. A., J. F. Saunders, R. K. Jacobson, and K. West. (2020). How gender and race stereotypes impact the advancement of scholars in STEM: Professors' biased evaluations of physics and biology post-doctoral candidates. *Sex Roles*, 82 (3), pp. 127–141.

Foucault, M. (1990). *The history of sexuality volume 1: An introduction*, trans. Robert Hurley. New York: Vintage.

Fricker, M. (2007). *Epistemic injustice: Power and the ethics of knowing.* New York: Oxford University Press.

Fricker, M. (2012). Silence and institutional prejudice. In: S. L. Crasnow, and A. M. Superson (eds.), *Out from the shadows: Analytical feminist contributions to traditional philosophy*, Oxford: Oxford University Press, pp. 287–306.

Goldenberg, M. (2021). *Vaccine Hesitancy: Public Trust, Expertise, and the War on Science*. Pittsburgh: Pittsburgh University Press.

Grasswick, H. (2017). Epistemic injustice in science. In: I. J. Kidd, J. Medina, and G. Pohlhaus, Jr., (eds.), *The Routledge handbook of epistemic injustice*. New York: Routledge, pp. 313–323.

Hall, K. Q. (2017). Queer epistemology and epistemic injustice. In: I. J. Kidd, J. Medina, and G. Pohlhaus, Jr., (eds.), *The Routledge handbook of epistemic injustice*. New York: Routledge, pp. 158–166.

Ivy, V. [McKinnon, R.]. (2019). Gaslighting as epistemic violence: 'Allies,' mobbing, and complex posttraumatic stress disorder, including a case study of harassment of transgender women in sport. In: B. R. Sherman and S. Goguen (eds.), *Overcoming epistemic injustice: social and psychological perspectives*, London: Rowman & Littlefield International, pp. 285–302.

Jordan-Young, R. M., and K. Karkazis. (2019). *Testosterone: An unauthorized biography*. Cambridge, MA: Harvard University.

Khubchandani, J., and Y. Macias. (2021). COVID-19 vaccination hesitancy in Hispanics and African-Americans: A review and recommendations for practice. *Brain, Behavior, & Immunity-Health*, 15, 100277.

King, M. M., C. T. Bergstrom, S. J. Correll, J. Jacquet, and J. D. West. (2017). Men set their own cites high: Gender and self-citation across fields and over time. *Socius*, 3, 2378023117738903.

Lee, J. Y. (2021). Anticipatory epistemic injustice. *Social Epistemology*, 35 (6): 564–576.

Leuschner, A., and M. Fernandez Pinto. (2021). How dissent on gender bias in academia affects science and society: Learning from the case of climate change denial. *Philosophy of Science*, 88 (4), pp. 573–593.

Leuschner, A., and M. Fernandez Pinto. (2022). Exploring the limits of dissent: The case of shooting bias. *Synthese*, 200: 326.

Longino, H. E. (1990). *Science as social knowledge: Values and objectivity in scientific inquiry*. Princeton, NJ: Princeton University Press.

Lorde, A. (1984). The master's tools will never dismantle the master's house. Taken from *Sister outsider: Essays and speeches*. (2007). Berkeley, CA: Crossing Press, pp. 110–114.

May, V. M. (2014). Speaking into the void? Intersectionality critiques and epistemic backlash. *Hypatia: A Journal of Feminist Philosophy*, 29 (1), pp. 94–112.

Mayor, A. (2005). *Fossil legends of the first Americans*. Princeton: Princeton University Press.

Mayor, A. (2008). Suppression of Indigenous fossil knowledge: From Claverack, New York, 1705 to Agate Springs, Nebraska. In: R. Proctor and L. Schiebinger (eds.), *Agnotology: The making and unmaking of ignorance*. Stanford: Stanford University Press, pp. 163–182.

Medina, J. (2017). Varieties of hermeneutical injustice. In: I. J. Kidd, J. Medina, and G. Pohlhaus, Jr. (eds.), *The Routledge handbook of epistemic injustice*. New York: Routledge, pp. 41–52.

Moss-Racusin, C., J. Dovidio, V. Brescoll, M. Graham, and J. Handelsman. Science faculty's subtle gender biases favor male students. (2012). *Proceedings of the National Academy of Sciences*, 109 (4) pp. 16474–16479.

NSF (National Science Foundation), National Center for Science and Engineering Statistics, (2023). *Diversity and STEM: Women, minorities, and persons with disabilities in science and engineering*. https://ncses.nsf.gov/pubs/nsf23315/report.

Pohlhaus, Jr., G. (2012). Relational knowing and epistemic injustice: Toward a theory of willful hermeneutical ignorance. *Hypatia: A Journal of Feminist Philosophy*, 27 (4): 715–735.

Pohlhaus, Jr., G. (2017). Varieties of epistemic injustice. In: I. J. Kidd, J. Medina, and G. Pohlhaus, Jr., (eds.), *The Routledge handbook of epistemic injustice*. New York: Routledge, pp. 13–26.

Spivak, G. C. (1988). Can the subaltern speak? In: C. Nelson and L. Grossberg (eds.) *Marxism and the interpretation of culture*. Chicago: University of Illinois Press, pp. 271–313.

Woolford, A. (2015). *This benevolent experiment: Indigenous boarding schools, genocide, and redress in Canada and the United States.* Lincoln, NE: University of Nebraska Press.

Yount, L. (2006). *Modern astronomy: Expanding the universe.* Infobase Publishing.

7

HOW CAN WE KNOW BETTER?

Overview

In this concluding chapter, we summarize the contributions of feminist epistemology and philosophy of science and consider what normative recommendations for our epistemic and scientific practices follow from the various analyses they offer. To do so we characterize what it might mean to know "better" and identify six themes that have emerged in the previous chapters for advancing the aims of knowledge production in at least one of these senses.

How can we know better?

Each of the chapters of the book has addressed a question about knowledge and the responses offered by feminist epistemologists and philosophers of science. We have contrasted those responses with some of the approaches of traditional epistemology and considered some of their strengths and weaknesses. Feminists have not been the only critics of traditional epistemology and philosophy of science, but they have contributed to and built upon changes in the understanding of knowledge production that have gained momentum over the past half century. While at one time it was considered heresy to suggest that values might play a role in science, most philosophers of science now believe that values not only influence what we know, but that in some cases this might even be beneficial for science. Feminists have developed this idea in a variety of ways that have enriched the discussion. As we have seen, not all philosophers or even all feminist philosophers agree about

DOI: 10.4324/9781032693781-7

what role, if any, values ought to play in knowledge production. This is an ongoing discussion that feminists continue to participate in.

Another area in which feminist epistemologists have made contributions is social epistemology. Feminist understanding of knowledge production as social and the various ways in which the social locations of knowers can affect their epistemic participation is clearly consistent with mainstream social epistemology. Nonetheless, insights of feminist epistemology involving the differential distribution of power along various lines, such as gender, sexuality, race, class, and ability, and the effect that has on key components of knowledge production have not always been acknowledged or incorporated into the social epistemology literature. Feminist epistemology and philosophy of science has much broader acceptance than it once did, but it still is not always fully recognized as part of mainstream epistemology or philosophy of science.

In this final chapter, we offer an account of how the work of feminists in epistemology and philosophy of science makes implicit and explicit recommendations for improving knowledge production. This includes recommendations for how we *think* about knowledge and concepts related to knowledge production and recommendations for scientific *practice*. The previous chapters have illustrated this in a variety of ways, but here we offer a summary and clarification of what this means.

When we consider potential recommendations for improving knowledge, the question arises: What counts as "better knowledge production"? "Better" is ambiguous because there are different possible goals of knowledge and thus "better" might be understood in different ways. First, "knowing better" might mean "knowing more" and, from the previous chapters we have seen, this is indeed one way that feminist epistemology and philosophy of science have understood knowing better. When women are left out as knowers or as subjects of knowledge, we lack knowledge that is desirable to have. We know better when we know about women's health, women's daily lives, and women's interests.

Knowing better can also mean producing knowledge that is more objective, more accurate (either in terms of depth or breadth), or more reliable. Several of the feminist revisions of our understanding of objectivity seek to support the idea that knowledge production that is guided by feminist values is more objective. We see this for example with Helen Longino's contextual empiricism and with Sandra Harding's notion of strong objectivity (Chapter 4).

Knowing better might also be understood as knowing more justly. This might involve ensuring that venues of knowledge production are equally open to all and that the practices of knowledge production reflect the needs and interests of all who might be affected by the knowledge produced. The exploration of epistemic injustice in Chapter 6 emphasizes this way of thinking about knowing better and illustrates the entanglement of ethical with epistemic issues. Knowing justly might also mean prioritizing areas of knowledge production that serve the interests of groups that are already disadvantaged, or who have had their interests historically neglected. For example, it might mean prioritizing aspects of climate change research that would help those who are disproportionately impacted by the threats of climate change and who have historically benefited less from, or been less responsible for, the production of CO_2 emissions.

Another way that we might understand "knowing better" is that we know better when we have a better understanding of the practices and processes through which knowledge is produced. This is one of the clearest lessons from feminist epistemology and philosophy of science and each of the other ways of knowing better follows from it, since each depends on understanding what might be missing from past epistemic practices.

The lessons we can draw from work in feminist epistemology and philosophy of science discussed throughout this book point to recommendations for how we might know better in each of these four senses. We summarize those lessons and how they might contribute to better knowledge production (in some sense) through the following six themes:

- What is known and who knows are intertwined.
- We depend on others for knowledge.
- Diversity plays important roles for epistemic communities.
- Better knowledge might involve feminist values.
- Science can be objective (but maybe not how we thought).
- Epistemology and ethics are interconnected.

In what follows, we discuss each of these indicating how they contribute to knowing better in the various ways we have identified and consequently how the insights from feminist epistemology and philosophy of science enable better knowledge production.

What we know and who knows are intertwined

Chapter 2 explored how who knows and what is known mutually affect each other. Feminist epistemologies challenge the assumption that knowers are isolated individuals and, instead, identify knowers as social beings both in the sense of being socially situated and in their reliance on others. In shifting to understanding knowers as social, feminist epistemology also directs attention to the role of power relations in the production of knowledge. In Chapter 6, we explored how the testimony of members of marginalized groups can be discounted, ignored, or difficult for members of such groups to express, either because of direct suppression of their views, willful or accidental negligence, or due to the absence of appropriate conceptual resources. In Chapter 5, we discussed how power dynamics can focus attention on some topics and lead to ignorance of others. Chapter 3 considered how the objects of inquiry are built out of the objects of the world insofar as they are of interest to the knowers. When the interests of some are discounted or subordinated to the interests of others, either intentionally or unintentionally, then gaps in knowledge result. When the differential interests of knowers are acknowledged, it opens the door to recognizing that the objects of world give rise to a variety of objects of inquiry relative to those differential interests. Knowers will ask different questions when their interests and goals vary. Consequently, when we evaluate knowledge and consider what counts as better knowledge we must ask "Better for whom and for which purposes?"

As we noted in Chapter 2, social location is complex. Women are not a homogeneous group but are socially located in a variety of ways – in terms of sexuality, (dis)ability, race, ethnicity, class, and other characteristics which intersect with gender. A full awareness of how knowers are situated thus leads to an intersectional analysis of knowers. At the same time, it complicates our understanding of knowers, since each individual may be socially situated across a variety of categories, all of which may be salient in different ways to their access to knowledge and their ability to participate in knowledge production. Kristie Dotson and Marita Gilbert offer the following characterization:

> This understanding of complex social identities attempts to move away from understanding social identities according to one salient or foundational identifier, which often serves to render unintelligible the fact that we simultaneously belong to multiple communities with different social and political features *that travel and change through the traveling.*
>
> *(Dotson and Gilbert 2014: p. 876)*

The idea of "traveling" captures the instability and context dependence of social identities. This variability can be understood as a strength rather than a weakness. If we must always ask who knows, what interests shape knowledge, and why knowledge is sought, then we must also evaluate knowledge relative to those interests. But that does not commit us to a pernicious relativism, since the question of whether the knowledge produced is responsive to those interests is, as argued in Chapter 3, constrained by empirical, theoretical, and pragmatic considerations.

The recognition of knowers as socially situated is also tied to a key feature of feminist methodology. Feminist researchers, particularly in the social sciences, have called for reflexivity – paying attention to one's own social location as a researcher and consciously exploring the ways in which it can make a difference to one's research. Reflexivity involves having an awareness of how those being studied may have different social locations, and hence experiences and interests, than those of the researcher. Reflexivity also asks of researchers that they acknowledge their own social location and the assumptions inherent with being so located. This includes assumptions that are built into disciplinary vocabularies, methodologies, and practices. The recognition of the epistemic relevance of social location can be seen particularly clearly in the literature on standpoint theory and in the adoption of standpoint methodology among feminist researchers. For example, Sandra Harding's strong objectivity (see Chapter 4) depends on reflexive evaluation of the assumptions and methods researchers depend on.

Notice that, in our discussion of reflexivity, we are not claiming that only those who are members of a particular group are able to know about that group. Such an interpretation of standpoint theory is widespread and, in our view, mistaken. How reflexivity plays out is not always obvious. Often, it is taken to involve no more than merely identifying and acknowledging the researcher's relationship to the community that is being researched – in other words, acknowledging how the researcher is socially located. Such a minimal acknowledgment indicates some self-awareness but is unlikely to make an epistemic difference. Reflexive research should acknowledge social location but, more importantly, it must also examine its implications – specifically the implications of power on knowledge production.

A similar worry can be raised about paying lip service to intersectional analysis. Simply noting that all of us belong to myriad social categories at the same time, and thinking of that as intersectionality, eliminates the political implications inherent in Crenshaw's original analysis. Intersectional analysis examines the ways in which the intersecting matrices of power that social categories indicate produce oppression in ways that cannot simply be reduced to the oppression of any one category, nor captured by "adding up" the oppression of the various categories. To ask for an intersectional analysis is to ask for an investigation into how the interaction results in a new or different form of disadvantage. It involves exploring how the social institutions, structured on explicit or implicit assumptions about who people are, produce inequities that call for rectification. These reflections point to the more recent work in feminist epistemology and philosophy of science – work that investigates how and why some groups have not participated in knowledge production and how they might come to participate.

An awareness of how knowers and what is known are intertwined, supports knowing better when understood as knowing more since it offers one of many motivations for diversity, both within a knowing community and of knowing communities. Within a knowing community, the interests of those who lack power often differ from those who are dominant. Consequently, features of the world that matter to those who are in a position of power may not support the purposes and goals of those who are subordinated. As a result, the production of knowledge may not serve everyone. This potential inequity is motivation to increase diversity of knowers, and for promoting a plurality of knowing communities, since knowledge that serves one portion of the population may not serve others.

It seems paradoxical that knowing more about subjects (knowers) would also increase objectivity. As we have seen in Chapters 3 and 4, that this is so depends in part on rethinking our understanding of objectivity. Indeed, this is what the feminist epistemologists and philosophers of science we have discussed have proposed. Longino's call for diversity within the knowing community offers one way in which flawed assumptions may be uncovered. Harding's endorsement of feminist standpoint theory speaks to this as well. But in addition, the evaluation of objectivity as knowledge that is adequate to the objects of inquiry depends on recognizing that what the objects of inquiry are relies on the interests and values of those seeking knowledge of them. While awareness of the interconnection of objects of inquiry with knowers does not in itself lead to knowing that is more just, epistemic communities that are more socially just would value the interests, testimony, and participation of those from subordinated communities.

We depend on others for knowledge

Another theme of Chapter 2 is that knowledge is social, not only because knowers are socially situated but also because it is not produced by isolated individuals. Any one person's skills and experiences are limited and we must rely on the testimony of others, their ability to produce the data we need for knowing, and the results of their research in complementary areas of study. It is not only other individuals, but past practices and theories, institutions, organizations, and social networks that are also implicated in all future knowledge production. The differential distribution of power in society is evident both in such social institutions and the funding that supports them. Awareness of such forces brings attention to how they can skew interests away from those of marginalized and oppressed groups by determining what researchers focus on in inquiry as well as what they take to be relevant evidence. Again, this may result in gaps in knowledge or ignorance as we saw in Chapter 5. Addressing this may aid in reducing ignorance (producing more knowledge).

In addition, the fact that we depend on others to know, also demonstrates the importance of epistemic trust. We must often depend on the expertise or testimony of others and that dependence makes us vulnerable in certain ways. Ideally, we would trust those who are trustworthy or credible or reliable. But we can sometimes be mistaken about whether a person or group or source is trustworthy, and we can be let down or led astray. Knowing better, in the sense of knowing more reliably or accurately will require identifying and understanding the conditions under which epistemic trust is warranted or justified. Feminist scholars (in both philosophy of science and in ethics) have argued that warranted trust involves not only epistemic considerations (e.g., reliability, epistemic competency), but also ethical dimensions (e.g., Baier 1986; Scheman 2001; Almassi 2012; Frost-Arnold 2013). Epistemic trust may require evidence, for example, that our epistemic sources have our best interests at heart or care about our wellbeing (de Melo-Martín and Intemann 2018; Goldenberg 2021). It may also require evidence of moral reliability or qualities such as honesty and transparency. This, in turn, may have implications for the kinds of practices that epistemic or scientific communities should adopt – that is, practices that would promote honesty, transparency, and a commitment to the public interest in addition to demonstrating epistemic reliability and competency.

Furthermore, because we depend on others for knowing, if there are groups whose epistemic competency or reliability is discounted in ways that are not warranted or justified, that can negatively impact the production of knowledge and, as we saw in Chapter 6, lead to epistemic injustice. Knowing better, in the senses of knowing more accurately and more justly may require identifying the ways in which implicit bias or power differences might cause certain testimony to be devalued, ignored, or disregarded as relevant evidence. While more work is needed to identify and understand the mechanisms of various kinds of epistemic injustices, work by feminist, decolonialist, and race theorists is foundational to helping us begin to understand these phenomena.

As we saw with standpoint theory, one remedy for the types of flaws in knowledge production just mentioned is "starting from the lives of women" or more generally "starting from the lives of the oppressed". This means taking seriously the testimony of those who might otherwise be dismissed thus addressing issues of epistemic injustice, but at the same time creating an opportunity to fill gaps in knowledge and decrease ignorance. The awareness of how we depend on others for knowledge may also enable us to know more objectively – once again, understanding objectivity as it has been described by the feminist theorists discussed in previous chapters (although primarily in Chapter 4).

Diversity plays important roles for epistemic communities

An implicit recommendation that comes from the critique of the traditional knower is that epistemic communities need to be more diverse to produce better knowledge. As we have seen, there are several arguments for increasing the diversity within the community. First, there are equity arguments. A gender imbalance in any profession is a prima facie indication of discrimination: either a lack of access to the education required for entering that discipline or some unjustified discrimination in hiring in that discipline. Research on women in science, technology, engineering, and mathematics (STEM) disciplines seems to bear this out, but it also reveals that women often leave these disciplines even after they attain the appropriate education and have been hired for jobs. The "chilly climate" research discussed in Chapter 2 offers one explanation. Still another reason why women are prone to leave STEM disciplines is that responsibilities for children still fall more heavily on women than men and the institutional settings of knowledge production rarely provide adequate family leave or childcare options. Women in these fields often report that they did not have as many children as they would have liked, or did not have children at all, in part because of their work situation (Ecklund and Lincoln 2011).

A number of philosophers of science have explored the link between diversity and better knowledge production. Carla Fehr (2011), for example, considers one way of making an argument for diversity based on Helen Longino's contextual empiricism (discussed in Chapter 4). Recall that Longino provides four norms that must be met within a scientific community in order for transformative criticism to be possible: 1) publicly recognized forums for the criticism of evidence, methods, assumptions, and reasoning; 2) uptake of criticism, which involves altering beliefs and theories in response to the critical discourse; 3) publicly recognized standards for the evaluation of theories, hypotheses, and observational practices; and finally, 4) tempered equality of intellectual authority, which should ensure that power dynamics within the community do not silence marginalized voices. Using Longino's account, it can be argued that the diversity of the community improves the critical discourse. Fehr refers to this argument as a "diversity promotes excellence" argument.

Fehr argues that Longino's account of community and the social nature of knowledge is not rich enough to fully support the diversity promotes excellence argument. She notes that it is quite possible for better knowledge to result from the diversity of the broader community, without fully diversifying the scientific community. Scientists might become aware of marginalized alternative points of view and, in doing so, uncover problematic assumptions without increasing the actual diversity of the formal knowledge producing community. Fehr refers to this as "diversity freeriding", since there is a sense that under these circumstances those who are doing the "diversity work" need not be part of the formal community of knowledge producers. In such circumstances, the argument would provide no support for redressing, for example, the disproportionately lower number of women who are full professors at universities.

Fehr proposes distinguishing between communities that are epistemically diverse and those that are situationally diverse. A situationally diverse community is one in which "membership consists of individuals with different social and material locations (gender, race, class, sexuality, etc.)" (Fehr 2011:146). But it is epistemic diversity that Fehr thinks should be promoted. In order for a community to be epistemically diverse its members must hold "a range of different background assumptions, and theoretical and methodological perspectives" (Fehr 2011:146). The mere presence of women or other marginalized groups in a research community may ensure situational diversity but it does not ensure epistemic diversity, since those who are members of marginalized groups may be subject to various forms of epistemic injustice that limit their full epistemic participation. While they may hold different background assumptions, they may not be heard when they voice them (testimonial injustice) or they may not even voice them if they are not in positions that carry equal authority to those of the dominant voices (testimonial smothering) (Dotson 2011) as discussed in Chapter 6. Thus, Fehr distinguishes between ineffective and effective epistemic diversity, where the latter requires an environment in which the full participation of all members of the research community is encouraged and supported.

Kristen Intemann (2010) has similarly explored the potential of Longino's account of science as social for supporting a call for diversity. She argues that while Longino seems to focus on the importance of diversity of interests and values in catching bias, situational diversity (or diversity of identities) may also be important within research communities to produce better knowledge. As we discussed in Chapter 4, Intemann's idea is that through combining standpoint theories with something like Longino's contextual empiricism, we get a feminist standpoint empiricism that incorporates the need for social diversity, as well as diversity of interests and values. This might address some of the challenges we saw in Chapters 3, 4, and 5, where we explored how particular interests (often unconsciously) shape what we know, the kinds of frameworks adopted, and result in ignorance about phenomena not relevant to those interests. By having communities that are socially diverse, it is also likely that more diverse experiences, values, and interests will also be represented in communities. As a result, it is more likely that a greater range of phenomena will be investigated and that a pluralism of methodologies, research questions and strategies, and directions will be pursued. In this sense, knowing better the role that diversity plays in how knowledge is produced might also lead us to knowing more (in depth and breadth) and knowing more objectively when communities are relevantly diverse.

The call for diversity to produce better knowledge is also directed at knowing more justly – in part by also knowing more about marginalized groups or the interests and issues that impact them disproportionately. Again, the inclusion of more points of view, including different values and interests, offers the possibility of a wider range of knowledge and knowledge that will serve a wider range of aims. This brings us to the role of values in knowledge production, which we address in the next section, and more generally to the intertwining of the ethical and the epistemic, which we explore later.

Better knowledge might involve feminist values

The importance of value diversity discussed in the last section brings us back to one of the themes that runs throughout our account of feminist epistemology and philosophy of science. Feminist critiques and analyses of knowledge production stem from the conviction that a commitment to feminist values improves knowledge production. What, exactly, are "feminist values"? While there might be different conceptions of what counts as a feminist value, or which feminist values might be beneficial to knowledge production, we have argued that a central feminist value is a commitment to economic and political gender equity. This minimally involves valuing knowledge that helps us to understand and overcome unjust inequalities and systems of oppression. Thus, the call for feminist values in the production of knowledge can be understood as a call for knowing more justly – as uncovering or revealing knowledge that will assist in this emancipatory project.

In Chapter 4, we saw that advancing this position involves challenging the value-free ideal (VFI) of science and knowledge. After all, if science is ideally value free, then it is not clear how feminist values could play a role in ideal knowledge production. While many philosophers of science have challenged the VFI (e.g., Rudner 1953; Kuhn 1977; Laudan 1984; Douglas 2000, Elliott 2017) feminists played an early and distinct role in doing so. For feminists, the VFI was problematic precisely because it involved deeply embedded assumptions that feminist scholars in the 1970s and 1980s found to be problematic. Proponents of early versions of the VFI implicitly assumed that scientists (or knowers) are individuals who can strip away all of their values and personal characteristics. They drew sharp distinctions between scientific investigation and reasoning and the social context in which it occurred. Moreover, women scientists entering the field saw their work dismissed in part because it appeared to contradict these assumptions. Thus, feminists were motivated to challenge the VFI to show that social features and context (including the social position of knowers and contextual values) were relevant to knowledge production. As we saw in Chapter 3, feminist work revealed that non-epistemic or contextual values play a role in what is studied, the way that research questions are framed, and how the objects of study are conceived. Values (including gendered values and norms) can play a role in our conceptual frameworks, descriptive language, and the ontologies we employ to classify objects of study. In Chapter 4 and 6, we also saw that values can play an important role in how we know, for example, in the selection of methodologies or in weighing the risks and possible errors that can result from uncertainties in drawing conclusions or interpreting data. In

Chapter 5, we also saw how values and interests can shape what we don't know as well as what we do – creating gaps in what we have knowledge about.

Yet, if values play an important role in the production of knowledge, in what sense might feminist values help produce better knowledge? This is a particularly important question because feminists also wanted to criticize sexist values or androcentric biases in science. In Chapter 3, we saw how this is related to the apparent bias paradox: if sexist or androcentric values are bad for science or objectivity because they *are* values, then how could *feminist* values possibly be any less biased or problematic?

As we saw in Chapter 4, some feminists have argued that feminist values themselves are more justified or more empirically adequate, in which case they could lead to more accurate or reliable knowledge. Others, such as Janet Kourany (2010), have argued that feminist values will help produce better knowledge in the sense that they will allow us to know more – and particularly know more about marginalized groups – precisely because they have been historically neglected. Thus, feminist values might correct historical imbalances that led to knowledge gaps and further a more comprehensive understanding of the world.

Values may also govern judgments about who we take to be full epistemic agents or who we take to have expertise. They may be involved in decisions about who to trust or whose testimony might be credible. In that sense, feminist values – and specifically a commitment to correcting implicit biases may also contribute to knowing justly and avoiding epistemic injustices.

In summary, recognizing that values affect each phase of scientific inquiry is a first step toward scrutinizing the appropriateness of their roles throughout inquiry and better understanding the processes of knowledge production.

Science can be objective (but maybe not how we thought)

As we have seen, feminist epistemology and philosophy of science challenge traditional understandings of objectivity. This is both because they challenge the VFI and because, as Briana Toole puts it, "the notion of objectivity itself is not value free" (Toole 2022: 7). The ideal of the objective knower (agent objectivity) as detached and unemotional characterizes that knower as having a "view from nowhere". But there is no view from nowhere if, as feminist theorists claim, knowledge is always situated. Understanding objectivity as a view from nowhere is also problematically misleading in that it disguises that the social location of that view is actually that of the dominant group rather than a neutral position. As such it carries with it the interests, values, and goals of the dominant group. Toole proposes instead that objectivity requires a "view from many places". We have seen similar claims in other feminist critiques – for example, Longino's call for tempered equality of intellectual authority and diversity in the epistemic community and Harding's strong objectivity together with standpoint theory's insistence on the importance of starting knowledge production from the lived experience of those who are marginalized. How do feminist understandings of objectivity lead to better knowledge?

An example of how feminist epistemology can improve knowledge production can be seen in some recent feminist work in data science. Data science:

> is a newly emerged research domain that includes several types of expertise relevant to data analysis, such as computational and statistical skills; epistemological expertise, including an understanding of where data fit in the processes of knowledge production; and expertise on data governance and ethics.
>
> *(Beaulieu and Leonelli 2022: 219)*

Data science and quantitative research more generally are interesting sites for thinking about feminist alternative conceptions of objectivity because of the widespread understanding of data, quantitative information, and the use of numbers in general as a way of achieving objectivity. Because objective knowledge is thought to be a desirable goal and because these methods are so powerful, they have come to dominate research in nearly all arenas of knowledge production. As Anne Beaulieu and Sabina Leonelli put it, "Most contemporary societies privilege ways of knowing that are grounded on data" (Beaulieu and Leonelli 2022: 3). One of the reasons for this is the idea that the objectivity of the data appears to derive from the detachment that numbers bring – they appear to strengthen agent objectivity. In addition, the mathematical procedures through which quantitative data are manipulated to produce statistical conclusions appear to have authority because they are formal. Quantitative methodologies appear to enhance methodological objectivity. In addition, because numbers can be stripped of context, they seem to be easy to transport from the venue in which they were produced to other sites and other uses.

But the sort of critique that feminists have offered of traditional understandings of objectivity can be applied to data science as well. To see why, it is useful to consider what steps are involved in producing, reporting, and using data. Beaulieu and Leonelli discuss what they call "data journeys". They use the term to "designate the movement of data from their site of production to many other sites where they are processed, mobilized and repurposed" (2022: 218). Their analysis illustrates how each stage of the journey involves decisions that affect the data and have implications for conclusions that may be drawn from them. As we saw in Chapter 3, decisions about what is significant, shape what it is that we know and, as we saw in

Chapter 4, decision points throughout the process of knowledge production provide moments when interests and values are incorporated into knowledge.

In their book *Data Feminism*, Catherine D'Ignazio and Lauren Klein (2020) offer an intersectional feminist analysis of how such decisions about what is significant affect data science practices – how data are collected, interpreted, and used. As an example, they look at Serena Williams's life-threatening experience giving birth in 2017 and identify it as a moment in which an awareness of the high rate of maternal mortality in the USA first came to public attention. Why hadn't data on maternal mortality been reported prior to this time? D'Ignazio and Klein tell us that "Nobody was counting. A 2014 United Nations report, coauthored by SisterSong, described the state of data collection on maternal mortality in the United States as 'particularly weak'" (D'Ignazio and Klein 2020: 23). Unexamined decisions about what is important to count can result in a failure to count what is significant for women and other marginalized groups, as we saw in the case of women and heart disease. Such failures result in ignorance (discussed in Chapter 5) that may hinder the ability to serve social justice goals or may even be harmful. That *Pro Publica* only began reporting data on maternal mortality in 2017 indicates that the information was not deemed significant to collect. Once it was collected and reported, it was possible to see that maternal mortality in the US is not commensurate with those of other advanced industrial nations. Furthermore, data shows that maternal mortality has steadily increased in all groups since it was first collected. A closer examination shows that Black women are more likely to die in childbirth than White women.[1]

As a remedy for such harmful lacunae in data, D'Ignazio and Klein propose explicitly adopting feminist principles to revamp practices surrounding the collection, interpretation, communication, and use of data to better serve emancipatory goals. These feminist principles include the recognition of knowers as always situated, examination of the assumptions that guide knowledge production (including assumptions about what is significant), and attention to context. Through these practices they aim to achieve a form of feminist objectivity – they mention both Harding's strong objectivity and Donna Haraway's partial perspectives (1988) – to replace what Ruha Benjamin (2019) has called "imagined objectivity" – the belief that data and associated algorithms are free from bias because they are distanced from the humans who produced them. Since the view from nowhere that this distance is supposed to provide is not possible, imagined objectivity disguises the biases of those who collect, produce, communicate, and use data – the majority of whom are cis White men as D'Ignazio and Klein remind us.

As part of their analysis D'Ignazio and Klein use Patricia Hill Collins's idea of matrices of domination – the institutions, organizations, and social structures that maintain power for the dominant groups. Identifying who has the power and then examining how that might affect data production is the first step to improving data science. Better data science also calls for diversity of social location to be represented among practitioners in order to reflect a view from many places instead of the impossible view from nowhere. Additionally, D'Ignazio and Klein urge transparency about who is collecting data, how they are collected, with particular attention to power relations, as well as transparency about the assumptions and circumstances under which data are cleaned, interpreted, communicated, and used (Harding's strong objectivity).

D'Ignazio and Klein are not alone in thinking that the practices surrounding quantificational techniques that so dominate current knowledge production are in need of revamping. The close association of quantification with traditional objectivity, coupled with practices that have biased the use of those methods in favor of the status quo, have resulted in the perception that feminism and quantitative methods conflict with each other. Yet many feminist social scientists do use quantitative techniques. Those who do have often had to defend their methods to both other feminists and non-feminist colleagues who reject the idea that sexist, racist, or other biases may be creeping into their work. While feminist social scientists do not all share the same views on objectivity, much of their work supports non-traditional understandings of objectivity that are more in line with those discussed in Chapters 3 and 4. For example, geographer Mei-Po Kwan specifically rejects the traditional notion of objectivity, writing, "Feminist objectivity should be understood in terms of the situated knowledges based on particular 'standpoints' or limited 'positions' of women's lived experiences in particular social and geographical context" (Kwan 2002 162). Kwan further admonishes against using quantitative methods as a license to make universal claims – again rejecting traditional ideals of knowledge. She also expresses concerns about the limitations of coding by gender, which may miss relevant nuances that could be picked up by more fine-grained coding in terms of gender by race, class, or other characteristics. She notes as well that existing measures need to be examined for potential biases and not used without reflection. Again, each of these worries illustrates the reflexivity with which feminists have argued research should be done.

While political scientists Katelyn E. Stauffer and Diana Z. O'Brien also acknowledge that quantitative methods are often associated with treating knowledge as value free and universalist, they argue that when used in conjunction with feminist methodology quantitative methods can improve knowledge production in political science (Stauffer and O'Brien 2018). They specifically mention how feminist methods can improve data production through focusing attention on the decisions those producing data make and the effects they may have on the results of research. They use the example of survey data, noting that questions that seek information about gender may be subject to social desirability bias.[2] A feminist approach calls for researchers to be conscious of the way their social location affects the knowledge produced and so can guard against such bias skewing results. Questions of measurement also benefit from feminist scrutiny. As we saw in Chapter 3, how objects of inquiry are conceptualized alters the way that they are measured. This was clear in the example of democracy. If a key feature of democracy – universal suffrage – is understood as adult males having the vote rather than all adults having the vote, then how we measure democracy is altered. We also saw this in Sari van Anders's Sexual Configuration Theory (SCT), an approach that allows for measurement of aspects of sexuality that other measures fail to capture.

Although we have focused on objectivity in each of these examples, the main lesson has been that objectivity is improved through a better understanding of how knowledge is produced. One of the key insights of feminist approaches to knowledge production is the attention to context throughout the process. The relevance of context increases the complexity of the analysis of knowledge production. Feminist analyses of objectivity reflect this increased complexity by directing us to look at the various phases of knowledge production and scrutinize the decisions made at those stages – the interests addressed and served and the background of beliefs and values against which those decisions are made.

Epistemology and ethics are interconnected

At the beginning of this book, we started from the recognition that many scholars were historically confused by the very notion of "feminist epistemology" or "feminist philosophy of science". In Chapter 1, we discussed how feminism is typically understood to be a political movement, committed to a foundational ethical principle: that women ought to have equal social, political, and economic rights and opportunities to men. Feminism, like other ethical or political perspectives, offers goals and prescriptions for how the world *ought* to be. Epistemology and science, on the other hand, are projects aimed at understanding how the world *is*. Yet, we have seen throughout this book, and even in many of the other lessons discussed above, that the fields of ethics and epistemology are more intertwined than previously thought.

As we saw in Chapter 3, science is concerned not only with achieving true or accurate claims about the world, but particularly interesting or significant truths that are guided by our interests. Ultimately, science is an enterprise with limited resources and so decisions need to be made about where to best expend those resources. Such decisions depend on values and interests, such as judgments about which problems are the most important for science to resolve, and whose problems should be addressed. Of course, some areas of science might also be directed towards things that may not directly impact human wellbeing, but, even the judgment that those other areas are also important and valuable, is indeed a value judgment.

As noted above, and as we saw in Chapter 4, ethical values may also play necessary and important roles in decisions throughout the phases of research. Not only are they relevant to which areas of study have priority, but also the ways in which they are studied, judgments about what counts as good evidence, decisions about which sorts of risks of error are acceptable, and how knowledge should be used. Insofar as this is correct, this also raises additional ethical issues, such as which values ought to be endorsed and who should decide.

Because we depend on others for knowledge, as noted above, we must also engage in relationships of epistemic trust, which have been shown to involve ethical as well as epistemic considerations. Having warranted trust in the testimony or expertise of others may involve having reasons to believe they are honest, transparent, and have your best interests or welfare in mind. Yet, there is more work needed to think about what specific practices might cultivate those virtues within epistemic communities and the circumstances under which trust should be granted. This is another area where epistemology intersects with ethics.

There are also ways in which decisions about whose testimony to trust, or to whom we should grant epistemic authority, involve ethical considerations about who deserves to be treated with respect or whether groups are fully human. These are not just questions about who is reliable, but also about who should be included and whether their representation matters within epistemic communities.

Conclusions

In this final chapter, we have shown how feminist approaches can improve knowledge production in a variety of ways. Additionally, we have offered some examples of how they have done so in practice.

Feminist philosophy of science and epistemology have influenced feminist research in a variety of disciplines, but most particularly in the social sciences. In recent years, these areas of feminist philosophy have expanded with a new awareness of the importance of intersectional analysis but continued work is needed. The themes that we identify as running throughout this book raise additional questions. For example, given that we rely on others to know, more work is needed to clarify the roles that individuals play in epistemic communities, the kinds of diversity that might be valuable to epistemic communities, and how inclusivity and equity within epistemic communities is best promoted or achieved. When is ignorance epistemically problematic and what are the most promising strategies for addressing it? What are the multitude of ways that epistemic harms and epistemic injustices occur and how can these be remedied or prevented? Moreover, as scientific research becomes increasingly complex and specialized, how can we promote epistemic trust and how can feminist practices be utilized in emerging contemporary scientific research?

As we noted at the outset, there are many changes that have taken place in the world of non-feminist philosophy of science and epistemology as well. There is a greater awareness of the role of values and the importance of context in understanding how knowledge production works. Although not always fully acknowledged, feminist work has both contributed to and benefited from these changes. It is our hope that the account that we have offered here will provide a better understanding and appreciation of feminist philosophy of science and epistemology.

Discussion questions

1. What are some ways that we can understand "better knowledge production"? Can you think of any other meanings than those mentioned in this chapter?
2. What are some way in which what we know and who knows are interrelated? What does this have to do with objectivity of knowledge?
3. In *Data Feminism*, the authors discuss a facial recognition algorithm that was unable to recognize the face of a researcher who was studying it. What might be some reasons that the algorithm failed to recognize her (although it recognized a face that she drew on the palm of her hand)?
4. Discuss some ways that ethics and epistemology are connected? Are there examples that you are aware of that are not mentioned in the text?

Additional suggested readings

Cartwright, N. and R. Runhardt. (2014). Measurement. In: N. Cartwright and E, Montuschi (eds.) *Philosophy of social science*. Oxford: Oxford University Press, 265–287. An account of what goes in to creating good measures.

Code, L. (2006). *Ecological thinking: The politics of epistemic location*. New York: Oxford University Press. A discussion of other factors that lead better knowledge production.

Crasnow, S. (2021). Feminist methodology in the social sciences. In: S. Crasnow and K. Intemann (eds.) *The Routledge handbook of feminist philosophy of science*. New York: Routledge, pp. 368–380. A discussion of various feminist methodologies and debates about the feminist use of quantitative methods.

Leonelli, S. and N. Tempini (eds.) (2020). *Data journeys in the sciences*. Springer. An account of how and when data is able to travel and when it is not.

O'Connor, C. (2019). *The origins of unfairness: Social categories and cultural evolution*. Oxford: Oxford University Press. A feminist inspired use of game theory to explore social and economic inequality.

Notes

1 The CDC reports that Black women are 3 times more likely to die in childbirth than White women April 6, 2022. https://www.cdc.gov/healthequity/features/maternal-mortality/index.html#:~:text=Racial%20Disparities%20Exist,structural%20racism%2C%20and%20implicit%20bias.
2 Social desirability bias occurs when respondents believe that if they answer truthfully their answers will be disapproved of by the researcher or the public more generally. Consequently, they adjust their answers. For example, if a survey question asks whether gender would matter for a decision about who to hire, the respondent may actually believe that it does matter but might respond that it does not because of their awareness of the general social condemnation of gender discrimination.

References

Almassi, B. (2012). Climate change, epistemic trust, and expert trustworthiness. *Ethics & the Environment*, 17 (2), pp. 29–49.

Baier, A. (1986). Trust and antitrust. *Ethics*, 96 (2), pp. 231–260.

Beaulieu, A., and S. Leonelli. (2022). *Data and society: A critical introduction*. London: Sage Publications.

Benjamin, R. (2019). *Race after technology: Abolitionist tools for the new Jim code*. Medford, MA: Polity.

de Melo-Martín, I., and K. Intemann. (2018). *The fight against doubt: How to bridge the gap between scientists and the public*. Oxford University Press.

D'Ignazio, C., and L. F. Klein. (2020). *Data Feminism*. Cambridge, MA: MIT Press.

Douglas, H. (2000). Inductive risk and values in science. *Philosophy of Science*, 67 (4), pp. 559–579.

Dotson, K. (2011). Tracking epistemic violence, tracking practices of silencing. *Hypatia: A Journal of Feminist Philosophy*, 26 (2), pp. 236–257.

Dotson, K., and M. Gilbert. (2014). Curious disappearances: Affectability, imbalances and process-based invisibility. *Hypatia: A Journal of Feminist Philosophy*, 29 (4), pp. 873–888.

Ecklund, E. H., and A. E. Lincoln. (2011). Scientists want more children. *PLoS ONE*, 6 (8), e22590. https://doi.org/10.1371/journal.pone.0022590.

Elliott, K. C. (2017). *A tapestry of values: An introduction to values in science*. Oxford University Press.

Fehr, C. (2011). What's in it for me? The benefits of diversity in scientific communities. In: H. Grasswick (ed.), *Feminist epistemology and philosophy of science: Power in knowledge*. Dordrecht: Springer, pp. 133–155.

Frost-Arnold, K. (2013). Moral trust and scientific collaboration. *Studies in History and Philosophy of Science, Part A*, 44 (3), pp. 301–310.

Goldenberg, M. J. (2021). *Vaccine hesitancy: Public trust, expertise, and the war on science.* University of Pittsburgh Press.

Haraway, D. (1988). Situated knowledges: The science question in feminism and the privilege of partial perspective. *Feminist Studies*, 14 (3), pp. 575–599.

Intemann, K. (2010). Twenty-five years of feminist empiricism and standpoint theory: Where are we now? *Hypatia: A Journal of Feminist Philosophy*, 25 (4), pp. 778–796.

Kourany, J. (2010). *Philosophy of science after feminism.* Oxford: Oxford University Press.

Kuhn, T. S. (1977). Objectivity, value judgment, and theory choice. *The Essential Tension: Selected Studies in Scientific Tradition and Change.* Chicago: University of Chigaco Press, pp. 320–329.

Kwan, M. (2002). Quantitative methods and feminist geographic research. In: P. Moss (ed.) *Feminist geography in practice: Research and methods.* Oxford: Blackwell Publishers, pp. 160–173.

Laudan, L. (1984). *Science and values: The aims of science and their role in scientific debate.* Berkeley, CA: University of California Press.

Rudner, R. (1953). The scientist qua scientist makes value judgments. *Philosophy of Science*, 20 (1), pp. 1–6.

Scheman, N. (2001). Epistemology resuscitated: Objectivity as trustworthiness. In: N. Tuana and S. Morgen (eds.), *Engendering rationalities.* Albany, NY: SUNY Press, pp. 23–52.

Stauffer, K. E., and D. Z. O'Brien. (2018). Quantitative methods and feminist political science. *Oxford Research encyclopedia of Politics.* Oxford: Oxford University Press. https://doi.org/10.1093/acrefore/9780190228637.013.210.

Toole, B. (2022). Objectivity in feminist epistemology. *Philosophy compass*, 17 (11), e12885. https://doi.org/10.1111/phc3.12885.

INDEX

Foucault, Michel 111
Frazier, Darnella 35
Fricker, Miranda 103, 105, 106, 110, 111
Frye, Marilyn 83, 85

"gap feminist empiricism" 74
gender 3, 12, 21, 29, 58, 66, 67, 109, 135;
  discrimination 139n2; imbalance 128;
  issues 68; norms 4; and phases of
  research 60–64; and race biases 110;
  and sex differences 21; stereotypes 20,
  47, 48, 62; violence 81
gendered biases 47, 59
gendered metaphors 46
"gender valence" 47
Gero, Joan 49
Gilbert, Marita 125
Grasswick, Heidi 108, 110, 113

Hacking, Ian 36
Hancock, Ange-Marie. 31n5
Haraway, Donna 10, 22, 134
Harding, Sandra 12, 18, 20, 57, 64, 70–72,
  74, 75, 91, 123, 126, 127, 132, 134
"herd immunity" 93
"hermeneutical death" 111
hermeneutical injustice 110, 111, 112, 113
hermeneutical injustices 110–112, 117
heterosexual households 66
heterosexual marriage 66
hormone contraceptives 84
human knowledge 7
human papilloma virus (HPV) vaccines 26
human wellbeing 136

identity-based testimonial injustices 106
ignorance 82–83; discrediting 92–94;
  disinformation 95; neglect 89–91;
  selective representation 95–96;
  suppression 91–92; varieties of 84–89
"imagined objectivity" 134
Indigenous fossil knowledge 92, 112
Indigenous knowledge 112
inductive risk 62–64, 114–115
industry-funded science 91
institutional racism 105
intellectual authority, equality of 68
Intemann, Kristen 26, 27, 69, 74, 130
interdisciplinary research 48
intersectional analysis 126
intersectional feminist analysis 134
intersectionality 23, 49
inversion thesis 26, 70, 74

Ivy, Veronica 107

Jordan-Young, Rebecca 47

Kant, Immanuel 7, 18
Karkazis, Katrina 47
Keller, Evelyn Fox 20, 46
Kitcher, Philip 41
Klein, Lauren 134, 135
knowing communities 126, 127
knowledge 92; claims 24; communities 27,
  44, 68; experiential 92; objective 40, 65,
  133; Platonic conception of 21;
  production 17, 23, 24, 27, 28, 29, 36,
  37, 45, 46, 51, 59, 64, 68, 117, 123, 124,
  127, 128, 129, 130, 131, 132, 133, 134,
  135, 136, 137; scientific 6, 19, 58;
  situated 21–27; as social 27–29; social
  situatedness of 5; theory of 7;
  traditional ecological 92
Kourany, Janet 67, 132
Kuhn, Thomas S. 5, 28, 73
Kukla, Quill 114, 115
Kwan, Mei-Po 135

law enforcement 36
Lawrence, Stephen 105
Lee, Ji-Young 108
Leonelli, Sabina 133
Lewinsky, Monica 81
"linguistic terrorism" 111
Lloyd, Elisabeth 37
lock-key model 42
Longino, Helen E. 57, 65, 67, 68, 69, 73–
  74 110, 123, 127, 129, 130, 132
Lorde, Audre 111
"loving ignorance" 88
Lugones, Maria 88

"male" hormone testosterone 45
Maney, Donna 57
marginalized social group 26
Martin, Emily 46
masculine traits 45
maternal mortality 134
May, Vivian 104
Mayor, Adriene 89, 91, 112
McClintock, Barbara 20
Medina, José 111
Mendoza, Brenny 91
Mercator projection map 43
methodological objectivity 37, 38, 39,
  58, 75